Encyclopaedia of

Sustainable Water Resource Management

Encyclopaedia of

Sustainable Water Resource Management

Volume 1

Editor:
R.N. Sahay

AGROTECH PRESS
JAIPUR ▪ NEW DELHI

AGROTECH PRESS

Regd. Off. "Bony Residency", Gate No. 2
Opp. Tilak Public School, Vishweswariya Nagar,
Gopalpura Road, Jaipur - 302018, India
Ph: 0141 - 2761280, Telefax: 2761381
E-mail : agrotechpress@gmail.com

Delhi Office: 102, 1st Floor, Satyam House,
4327/3, Ansari Road, Darya Ganj,
New Delhi - 110002, India
Ph: 011-45652440
visit us: www.agrotechpress.com

Distributed By
KEYA INTERNATIONAL PRESS LLP
E-mail: keyainternationalpressllp@gmail.com

Published in India, 2025

ISBN : 9789384568269 (5 Vols. Set)

Typeset by Thomson Press

Printed and bound in India by Thomson Press Ltd.,
New Delhi

Contents

Preface

Integrated Water Resources Management (IWRM) has been defined as "a coordinated, goal-directed process for controlling the development and use of river, lake, ocean, wetland, and other water assets.

There are great differences in water availability from region to region - from the extremes of deserts to tropical forests. In addition there is variability of supply through time as a result both of seasonal variation and inter-annual variation. All too often the magnitude of variability and the timing and duration of periods of high and low supply are not predictable; this equates to unreliability of the resource which poses great challenges to water managers in particular and to societies as a whole. Most developed countries have, in large measure, artificially overcome natural variability by supply-side infrastructure to assure reliable supply and reduce risks, albeit at high cost and often with negative impacts on the environment and sometimes on human health and livelihoods. Many less developed countries, and some developed countries, are now finding that supply-side solutions alone are not adequate to address the ever increasing demands from demographic, economic and climatic pressures; waste-water treatment, water recycling and demand management measures are being introduced to counter the challenges of inadequate supply. In addition to problems of water quantity there are also problems of water quality. Pollution of water sources is posing major problems for water users as well as for maintaining natural ecosystems. In many regions the availability of water in both quantity and quality is being severely affected by climate variability and climate change, with more or less precipitation in different regions and more extreme weather events. In many regions, too, demand is increasing as a result of population growth and other demographic changes and agricultural and industrial expansion following changes in consumption and production patterns. As a result some regions are now in a perpetual state of demand outstripping supply and in many more regions that is the case at critical times of the year or in years of low water availability.

Operationally, IWRM approaches involve applying knowledge from various disciplines as well as the insights from diverse stakeholders to devise and implement efficient, equitable and sustainable solutions to water and development problems. As such, IWRM is a comprehensive, participatory planning and implementation tool for managing and developing water resources in a way that balances social and economic needs, and that ensures the protection of ecosystems for future generations. Water's many different uses—for agriculture, for healthy ecosystems, for people and livelihoods—demands coordinated action. An IWRM approach is an open, flexible process, bringing together decision-makers across the various sectors that impact water resources, and bringing all stakeholders to the table to set policy and make sound, balanced decisions in response to specific water challenges faced. In putting the IWRM principle into practice, many countries have adopted an approach where regulatory decisions such as water allocation and pollution licensing are implemented at the scale of the river basin or catchment. This has been accompanied by the emergence of institutional arrangements for water resources management that based on hydrological boundaries. While most of these institutions can be grouped as River Basin Organisations some are specifically mandated with managing groundwater water aquifers and lakes basins. Cap-Net, a UNDP capacity development programme for sustainable water management developed a training manual on IWRM for River Basin Organisations and works with networks of local capacity builders around the world to assist water managers with the concept of using an IWRM approach on the ground.

This book is an important contribution to international, regional and national debates on managing water resources and on enhancing activism to make integrated water management a possibility. It will be of interest to professionals in the fields of resource management, rural development, environmental studies, engineering and participatory management, as also to aid development agencies and NGOs involved in the fields of water use and management, sustainable development and food security. It will also be of use to students of development studies and sociology.

—Editor

1

The Groundwater Resource of Water

Groundwater may be defined as subsurface water that occurs beneath the water table in soils and geologic forms that are fully saturated. It is an integral part of the hydrologic cycle, and any approach to groundwater problems should recognise this. Groundwater is not only an important natural resource but also an essential part of the natural environment. Much of the folklore and many of the widely held misconceptions about the nature of groundwater. For example, that it occurs in underground lakes, rivers, and veins and can be detected by listening for the noise are dispelled by Lehr and Pettyjohn (1975) in their description of groundwater and its flow patterns.

The global hydrologic cycle consists of the movement of water between the oceans and other surface water, the atmosphere, and the land. A part of the precipitation, either as rainfall or melting snow, infiltrates the ground and percolates down through the unsaturated soil, known 18 as the aerated zone, to the zone of saturation or the water-table level. An estimate of the water balance of the world, shows the relative volumes of water contained in each part of the hydrologic cycle. Only 2.7% of all the water on the planet is fresh, and of that only 0.36% is easily available to users.

It has been estimated that about 40,000 billion gallons per day (bgd) pass over the conterminous United States as water vapour. Approximately 10% of this, 4,200 bgd, is precipitated as rain, snow, sleet, or hail, equivalent to an average uniform annual rainfall of 30 inches nationwide. About two-thirds of the precipitation evapourates or is transpired by vegetation. The remaining 1,450 bgd accumulates in ground and surface waters, flows to the sea or across national boundaries, is consumed, or evaporates from reservoirs.

***Table:** Estimate of the Water Balance of the World*

Parameter	*Surface Area (km)x10*	*Volume (km)x10*	*Volume (%)*	*Equivalent Depth(m)**	*Residence Time*
Oceans and Seas	361	1370	94	2500	~4000 yrs
Lakes and Reservoirs	1.55	0.13	<0.01	0.25	~10 yrs
Swamps	<0.1	<0.01	<0.01	0.007	1-10 yrs
River Channels	<0.1	<0.01	<0.01	0.003	~2 wks
Soil Moisture	130	0.07	<0.01	0.13	2 weeks-1 yr
Ground water	130	60	4	120	2 wks-10,000 yrs
Icecaps and Glaciers	17.8	30	2	60	10-1000 yrs
Atmospheric Water	504	0.01	<0.01	0.025	~10 days
Biospheric Water	<0.1	<0.01	<0.01	0.001	~1 wk

*Computed as though storage were uniformly distributed over the entire surface of the earth

Source: R. Allan Freeze, John A. Cherry. *Groundwater* (Engelwood Cliffs, N.J.: Prentice Hall, Inc., 1979), p. 84. Reprinted by permission of the publisher.

Of the 1,450 bgd only 675 bgd are usually available for intensive beneficial uses. The ability of an aquifer to store and transmit water is a function of its permeability and porosity. When the saturated substratum is sufficiently permeable to store and transmit significant quantities of water, the geological formation is called an aquifer. There are two main types of aquifers, confined and unconfined. An unconfined or water-table aquifer contains water under atmospheric pressure; the upper 19 surface of the water is called the water table and may rise and fall according to the volume of water stored, which is dependent upon seasonal cycles of natural recharge. A perched aquifer is one in which a limited layer, or lens, of impermeable material occurs above the water table, forming a thin zone of saturation above it. It is a type of water-table aquifer. The second major category of aquifer is the confined, or artesian, aquifer.

These are bounded top and bottom by layers of relatively impermeable geologic formations termed aquitards, or confining layers. The aquifer is completely saturated with water that is under greater than atmospheric pressure. An artesian aquifer is not recharged everywhere uniformly, but in one or more general recharge areas. Water levels in nonpumping wells of unconfined aquifers correspond to the level of the water table and therefore vary according to the volume of water in storage. Water levels in nonpumping wells tapping confined aquifers are dependent upon the artesian pressure of the water in that aquifer, and in some cases the water may exceed the top of the well casing, thus causing a flowing well. The hypothetical projection of such water levels is known as the potentiometric surface. The aquitards, or confining layers, are not totally impermeable and permit some recharge or discharge to lower confined aquifers. Aquifers can occur in unconsolidated materials such as sand and gravel or in consolidated material or bedrock. The latter may consist of carbonate rock, volcanic rock, or fractured igneous, metamorphic, and sedimentary rocks. Sand and gravel aquifers usually contain the most groundwater, but high-yield wells can also occur in carbonate and volcanic rocks.

Groundwater Movement

Groundwater moves in response to gravity, pressure, and friction, the first two driving the water, the latter resisting motion. Due to the complexity of the channels through which groundwater flows, it is difficult to construct a model of the movement of groundwater at a microscopic level. The French engineer Darcy, however, formulated an empirical law in the mid-nineteenth century that effectively averages the microscopic complexities, providing a macroscopic model for groundwater movement. Darcy's Law relates the rate at which groundwater flows across a surface to the rate of change of energy of the groundwater along the flow path. Under ideal 20 homogeneous isotropic geologic groundwater conditions, the average linear groundwater velocity ((v)) can be expressed by the Darcy relation:

$$v = - [K\ dh/dl]/n$$

v = the average linear groundwater velocity

where:

K = hydraulic conductivity

dh/dl = hydraulic gradient

n = transport porosity

For both granular and fractured media, the same equation is used to compute values for average linear groundwater velocity. The great difference is in the value of *n*. For granular media n nearly represents the total porosity; in fractured rock or clay n represents the total void space in connected fractures within a unit volume of media. The value of n in granular material is usually between 0.3 and 0.45; in fractured material n is usually very small, of the order of 10-*f*"2 or 10-*f*"4. When the values of *K* and *dh/dl* are similar in granular and fractured media, the average linear groundwater velocity may be orders of magnitude larger in fractured media than in granular media. The velocity of groundwater in a horizontal sand and gravel aquifer is usually in the range of 0.05 to 1 meter per day.

Darcy's Law is central to the derivation of equations used to model the flow of groundwater. Mathematical models, based on the physics of groundwater flow and boundary conditions imposed by the groundwater basin in question, usually take the form of a boundary-value problem. Bear (1972) discusses the application of the three methods used for solving such boundary-value problems, namely analytical, analogue, and numerical. Analytical methods result in explicit mathematical expressions of the solution and are useful for problems where the governing equations may be simplified and the boundaries of the basins idealised.

Analogue models are scaled physical models usually involving the construction of electronic equipment to simulate groundwater flow. Numerical mathematical techniques are the basis for digital computer simulation of transient groundwater flow in aquifers. Digital simulation of groundwater flow requires an expertise in computer programming and its use is bound to digital computers. However, it is more flexible in its ability to handle irregular groundwater boundaries and variations of groundwater movement through time and space than are the other two modelling techniques. The two most widely used numerical methods for solving groundwater equations are the finite-difference and finite-element techniques.

The main difference between groundwater and surface water is that the movement of groundwater takes place very slowly. Mixing of groundwater is slow, in contrast to the good mixing potential in the turbulent flow of most surface water. This factor is extremely important when considering the fate of contaminants.

Occurrence and Natural Quality of Groundwater

Most aquifers occur within 2,500 feet of the surface of the land. They may be thick or thin, extensive or local, very near the land surface or at considerable depths. It has been estimated that 30% of the stream flow of the United States is supplied by groundwater that emerges as natural springs or other seepage. In certain areas in times of drought, the stream flow in lowflow months may be largely provided by groundwater. The interrelatedness of surface water and groundwater is further underlined by the fact that under certain conditions seepage from lakes, rivers, streams, reservoirs, and canals may recharge aquifers.

The quantity of groundwater in storage is much greater than the volume of surface water available in streams and lakes. It comprises more than 96% of all the fresh water in the United States, the remaining 4% occurring in lakes, rivers, and streams. Estimates of groundwater within 2,500 feet of the land surface in the conterminous United States range from 33 quadrillion to 59 quadrillion gallons to as much as 100 quadrillion gallons.

As a point of comparison, Lake Michigan contains 1.3 quadrillion gallons of water. Not all of the groundwater in storage is available for use, however, because some is bound to soil particles. The cost of extraction may also restrict its use in certain areas. It would be possible to drill a well with a yield sufficient for domestic use in almost every region of the country.

The country has been divided up into 10 groundwater regions. This is considered the best broad classification of the groundwater situation in the conterminous United States. The division is based on the types of aquifers.

1. The western-mountain ranges consist mainly of igneous, metamorphic, and consolidated sedimentary rocks. Most of the groundwater occurs in rock fractures. The large amount of precipitation that falls in this area recharges aquifers in adjacent regions.
2. Alluvial basins consist of valleys surrounded by mountains in the arid southwest. Water levels in many aquifers have declined over recent years due to the use of large volumes of groundwater for agricultural irrigation.
3. The Columbia Lava Plateau consists of lava flows and unconsolidated sediments. In some areas there are large supplies of groundwater.
4. The Colorado Plateau has high, dry plateaus of sedimentary shale and sandstone and has a scarcity of productive aquifers.
5. The High Plains are extensive and semi-arid to sub-humid. The unconsolidated sedimentary rocks form the Ogallala Aquifer. Due to extensive agricultural irrigation, withdrawals have greatly exceeded groundwater recharge in some areas.
6. The glaciated central lowlands consist of glacial drift, sand, and gravel forming major groundwater reservoirs. In some heavily populated or highly industrialised areas the groundwater has been overdeveloped or polluted.
7. The unglaciated Central Region is composed of horizontal consolidated sedimentary rocks, with limestone and sandstone formations providing the major aquifers. The yields are low to moderate and adequate for domestic supply but would not be sufficient for irrigated agriculture.
8. The unglaciated Appalachians have high-yielding limestone and sandstone aquifers, and lower yielding shale aquifers. The water table may occur at a considerable depth below the surface. This region has an abundance of surface water year round.
9. The glaciated Appalachians region is similar to the unglaciated Appalachians.
10. The Atlantic and Gulf Coast Plain has an abundance of groundwater and surface water. The geology consists of unconsolidated gravel, sand, silt and clay, and limestone. Supplies of groundwater may not be abundant in some areas of Texas, Louisiana, Mississippi, and Alabama. Florida has a prolific aquifer.

Unfortunately there is more than one classification system for dividing the country up into water regions. For the purpose of compiling and analysing water-resources data for both surface water and groundwater, the U.S. Water Resources Council divided the nation into 21 major water-resources regions, 18 within the conterminous United States, and the other 3 being Alaska, Hawaii, and the Caribbean area. These 21 regions are further subdivided into 106 subregions. The regional divisions are areas that contain either the drainage area of a major river or the combined drainage of a series of rivers.

The EPA used another classification system in compiling data on groundwater pollution for various regions of the United States. The EPA system most closely resembles that described by Lehr et al. (1976), which is summarised above. Along with the geological characteristics of the groundwater regions, climatic conditions in particular, precipitation and temperature have effects upon the quantity and quality of the groundwater available. Variations from average precipitation may cause droughts or floods. Water is lost from the land to the atmosphere by evaporation and transpiration, or evapotranspiration. When evapotranspiration exceeds precipitation, groundwater recharge by percolation does not occur. In the arid areas of the Southwest, for example, annual precipitation is less than 10 inches, and potential evaporation could exceed this 4 to 20 times. In such areas groundwater recharge occurs mainly in wet, or multiyear, cycles. In the eastern states, the annual precipitation exceeds the evaporation, thus providing surpluses that contribute to stream flow. Regional variations in temperature may also affect groundwater in ways unrelated to surface evaporation. For example, frozen ground does not permit infiltration of rain, thus causing floods and preventing groundwater recharge.

The United States Geological Survey (USGS) published summary appraisals of the nation's groundwater resources between 1974 and 1982 for each of the regions that were briefly outlined above in the U.S. Water Resources Council scheme. These geological survey professional papers list groundwater usage for each region by categories of use (e.g., municipal water supply and agricultural irrigation). They list the natural groundwater quality and any problems that may be encountered due to the presence of natural iron in the water or to its salt content and hardness. In addition, they list present and potential anthropogenic threats to the quality of groundwater. The National Water Summary 1984, also put together by the USGS, contains descriptions of the occurrence, use, and general quality of groundwater resources of each state. Each summary contains: (1) the physiographic, hydrologic, and geologic framework of the state's groundwater system; (2) a description of the principal aquifers; (3) groundwater withdrawals and water-level trends; (4) a description of the state's groundwater management program(s); (5) maps showing geographic distribution of the principal aquifers; figures illustrating the areal distribution of major groundwater withdrawals and trends in water levels; and selected references.

Table: *Summary Appraisals of the Nation's Groundwater Resources, by Region*

(A)	Ohio — R. M. Boyd, Jr., 1974.
(B)	Upper Mississippi — R. M. Boyd, Jr., 1975.
(C)	Upper Colorado — Don Price and Ted Arnow, 1974.
(D)	Rio Grande — S. W. West and W. L. Broadhurst, 1975.
(E)	California — H. E. Thomas and D. A. Phoenix, 1976.
(F)	Texas-Gulf — E. T. Baker, Jr. and J. R. Wall, 1976.
(G)	Great Basin — Thomas E. Eakin, Don Price, and J. R. Harrill, 1976.
(H)	Arkansas-White-Red — M. S. Bedinger and R. T. Sniegrocki, 1976.
(I)	Mid-Atlantic — Allen Sinnott and Elliott M. Cushing, 1978.
(J)	Great Lakes — William G. Weist, Jr., 1978.
(K)	Souris-Red-Rainy — Harold O. Reeder, 1978.
(L)	Tennessee — Ann Zurawski, 1978.
(M)	Hawaii — K. J. Takasaki, 1978.
(N)	Lower Mississippi — J. E. Terry and C. T. Bryant, 1979.
(O)	South Atlantic-Gulf — D. J. Cederstrom, E. H. Boswell, and G. R. Tarver, 1979.
(P)	Alaska — Chester Zenone and Gary S. Anderson, 1978.
(Q)	Missouri Basin — O. James Taylor, 1978.
(R)	Lower Colorado — E. S. Davidson, 1979.
(S)	Pacific Northwest — Bruce L. Foxworthy, 1979.
(T)	New England — Allen Sinnott, 1982.
(U)	Caribbean — Fernando Gómez-Gómez and James E. Heisel, 1980.

Source: U.S. Geological Survey Professional Paper 813. U.S. Government Printing Office.

The quality of groundwater is often described in terms of hardness and salinity. Hardness reflects its calcium and magnesium content and is usually expressed as the equivalent amount of calcium carbonate. It can be viewed as a measure of usefulness for domestic and industrial purposes.

Water quality, however, is usually defined in terms of the concentration of its chemical constituents. As water moves through the hydrologic cycle, it interacts with the atmosphere, soils, and subsurface geologic formations, all of which affect its chemical composition. Thus, there is a natural background level for the chemical content of the water, and this level varies regionally and may be subsequently augmented from industrial and domestic sources. Freeze and Cherry (1979) discuss the natural chemical content of groundwater in detail. Rainwater is saturated with oxygen, nitrogen, and carbon dioxide gases and is usually slightly acidic, having a pH of about 5.6. The acidity may be increased by industrial pollutants, namely, the oxides of sulfur and nitrogen. The more acid the rainwater, the more likely it is to react with the geologic materials with which it comes into contact.

Rainwater percolating through the soil may increase in acidity due to the biological processes that occur in that zone. Plant and microbial respiration produce carbon dioxide, which increases acidity. It would be possible for the percolating water to become supersaturated with carbon dioxide. Acidity may also be increased by products of decomposition such as humic and fulvic acids, nutrient uptake by roots, and nitrifying bacteria, but these are minor factors compared with the amount of carbon dioxide produced by respiration. Thus, the water undergoes a chemical change during its passage through the soil to the underlying water-bearing formations.

Table: *Simple Groundwater Classification Based on Total Dissolved Solids*

Category	***Total Dissolved Solids(mg/l or g/m)***
Fresh water	0-1000
Brackish water	100-10,000
Saline water	10,000-100,000
Brine water	More than 100,000

Source: Freeze and Cherry, 1979.

Table: *Classification Based on Dissolved Inorganic Constituents in Groundwater*

Major Constituents (greater than 5 mg/l)	
Bicarbonate	Silicon
Calcium	Sodium
Chloride	Sulfate
Magnesium	Carbonic acid
Minor Constituents (0.01-10.0 mg/l)	
Boron	Nitrate
Carbonate	Potassium
Fluoride	Strontium
Iron	
Trace Constituents (less than 0.1 mg/l)	
Aluminum	Molybdenum
Antimony	Nickel
Arsenic	Niobium
Barium	Phosphate

Beryllium	Platinum
Bismuth	Radium
Bromide	Rubidium
Cadmium	Ruthenium
Cerium	Scandium
Cesium	Selenium
Chromium	Silver
Cobalt	Thallium
Copper	Thorium
Gallium	Tin
Germanium	Titanium
Gold	Tungsten
Indium	Uranium
Iodide	Vanadium
Lanthanum	Ytterbium
Lead	Yttrium
Lithium	Zinc
Manganese	Zirconium

Source: S. N. Davis and R. J. M. DeWiest, Hydrogeology (New York: John Wiley and Sons, Inc., 1966). Reprinted by permission of the publisher.

In the saturated zone, the water reacts with the geologic formations, increasing the content of total dissolved solids. Thus, the chemical quality of groundwater depends both upon its age and the geological formations encountered in its flow history. Carbonate formations would increase the magnesium and calcium content and also the concentration of bicarbonate due to the dissolution of calcite ($CaCO_3$) and dolomite (Mgò$CaCO_3$). Aluminosilicate minerals would increase the concentrations of sodium, potassium, magnesium, calcium, and silicon hydroxide ($Si(OH)_4$) in the water.

In consolidated deposits consisting of minerals from a range of sedimentary, igneous, and metamorphic sources, the order of encounter of the groundwater with the different assemblages determines the chemical constituency at a particular point in time. Sulfate-bearing minerals such as gypsum and anhydrite, although they occur less frequently than the carbonate and crystalline formations, are characterised by high solubilities. Thus, in older groundwater that has encountered sulfur-bearing minerals, the sulfate anion may dominate the bicarbonate anion. In very deep, old groundwater, the chloride anion may dominate both the bicarbonate and sulfate ones due to the presence of readily soluble minerals such as halite (NaCl) and sylvite (KCl).

The U.S. Public Health Service (USPHS) standard for total dissolved solids (TDS) in drinking water is 500 ppm, although the 1974 Safe Drinking Water Act considers waters containing up to 10,000 ppm TDS as potential sources of drinking water. The four types of naturally occurring groundwater that often exceed 10,000 ppm TDS are connate water, intruded seawater, magmatic and geothermal water, and water affected by salt leaching and the products of evapotranspiration.

In many areas of the United States, groundwater can be used for drinking with no pretreatment, in which case it is said to be a raw resource. In other areas pretreatment to correct hardness or colour may be necessary.

Where contamination by pathogenic organisms is a potential threat, chlorination may be indicated. Water from some aquifers is unusable due to its salinity or to the presence of naturally occurring toxic substances such as arsenic or radionuclides. Most of the freshwater aquifers are underlain by brackish or saline aquifers, and in general, salinity may be said to increase with depth.

Use of Groundwater

Groundwater is a major natural resource in the United States and is often more easily available than surface water. It is estimated that more than 50% of the population uses groundwater as its primary source of drinking water. These estimates are from unpublished data for 1970 from the USGS. They indicate that groundwater delivered by community systems supplied 29% of the population, and an additional 19% had its own domestic wells.

Approximately 36% of the municipal public drinking-water supplies came from groundwater. Of the rural population, 95% was dependent upon groundwater for drinking purposes in 1970. Of the total groundwater usage, 60% was used for public supplies and 40% for rural supplies. EPA's National Statistical Assessment of Rural Water Conditions estimated that 72.3% of all major rural household supplies was extracted from groundwater. This dependence varied regionally, with 87.5% of north-central rural households and 56.7% of western households dependent upon groundwater for water supplies.

This table indicates that 97% of water withdrawn for the nation's domestic rural use is supplied by groundwater. It is clear that the states west of the Mississippi River, in the area where irrigated agriculture is prevalent, depend heavily on groundwater. Arkansas, Nebraska, Colorado, and Kansas use more than 90% of their groundwater for agricultural activities. Irrigation was the principal use of all groundwater withdrawn across the country in 1980. The more humid eastern portion of the country is far less dependent on groundwater. The western states, again, are the main users of groundwater.

The most recent figures for groundwater use are from 1980 USGS data. The total use of fresh groundwater showed an increase of 22% from 1970 to 1975 and an increase of only 7% from 1975 to 1980. This change in the rate of increased use may be attributable to the fact that high demand for groundwater may decrease the ease and efficiency of its withdrawal, and in some cases its quality.

Table: *Groundwater Withdrawals as a Percentage of Total Freshwater Withdrawals for Rural Domestic Supply*

State	*Percentage*	*State*	*Percentage*
Alabama	100	Montana	94
Alaska	99	Nebraska	100
Arizona	100	Nevada	94
Arkansas	100	New Hampshire	98
California	93	New Jersey	100
Colorado	36	New Mexico	97
Connecticut	100	New York	89
Delaware	100	North Carolina	100
District of Columbia	0	North Dakota	100
Florida	100	Ohio	90
Georgia	100	Oklahoma	83
Hawaii	90	Oregon	87
Idaho	96	Pennsylvania	100
Illinois	97	Rhode Island	100
Indiana	90	South Carolina	100
Iowa	100	South Dakota	94
Kansas	86	Tennessee	100
Kentucky	91	Texas	84
Louisiana	100	Utah	90
Maine	98	Vermont	85
Maryland	100	Virginia	100
Massachusetts	100	Washington	78
Michigan	100	West Virginia	95
Minnesota	100	Wisconsin	100
Mississippi	100	Wyoming	92
Missouri	74		
Total national percentage			*97*

Source: After Heath, 1985.

Escalated costs resulting from these factors and the cost of the water itself have perhaps influenced users, especially irrigators, to be more efficient with groundwater use. Actual water-use figures for 1975 and 1980 (including saline groundwater withdrawals and water used for the generation of thermoelectric power) are given.

***Table:** Historical Trends in U.S. Groundwater Use as a Percentage of Withdrawals: 1950-1980*

	1950	*1960*	*1970*	*1975*	*1980*
Total Fresh Groundwater Withdrawals (bgd)	34	50	68	82	88
Public Supplies (%)	12	13	14	13	14
Rural Supplies (%)	8	6	5	5	5
Irrigation (%)	62	68	66	69	68
Industry (%)	18	13	15	14	13

Sources: Murray and Reeves, 1972, 1977; Makichan and Kammerer, 1961; Soiley et al., 1983.

Note: May not total 100% due to rounding.

In an average American community, the average per-capita water consumption is approximately 159 gallons per day (Last, 1980). In an American home, depending on the nature of the residence and the climate, exterior residential use may range from 5% to more than 150% of the interior use, averaging about 75%. This is primarily for watering lawns. Personal use of water, excluding laundry, toilet flushing, and so forth, accounts for 40% of the internal residential use and averages out to 6.3 gallons per capita per day.

***Table:** U.S. Groundwater Use: 1975, 1980*

	1975	*1980*
Population served by groundwater for public supplies	64,700,000	73,700,000
Total groundwater withdrawn (mgd) for public supplies	11,000	12,000
Rural use of groundwater (mgd) Domestic	2,700	3,300
Livestock	1,200	1,200
Total withdrawn for irrigation (mgd) toirrigate a total of 54,000,000 acres (1975) and 63,125,000 acres (1980)	57,000	60,000
Groundwater withdrawals for industrial use (mgd) Fresh	11,000	10,000
Saline	980	930
Water used for generation of thermoelectric power	1,390	1,600
Total withdrawals of groundwater (mgd)	85,270	89,030

Sources: Murray and Reeves, 1977; Soiley et al., 1983.

The safe yield of a groundwater basin is defined as the amount of water that can be withdrawn annually without producing an undesired effect. Undesired results include depletion of the groundwater reserves, intrusion of water of an undesirable quality, contravention of water rights, the deterioration of the economic advantages of pumping, excessive depletion of stream flow, and land subsidence. Any withdrawal in excess of a safe yield is termed an overdraft. It has been suggested that the optimal yield would be a more useful concept and should be determined by an optimal groundwater-management scheme that includes and best meets economic and social objectives associated with the use of water. Under certain conditions, optimal yields may involve the mining of groundwater to depletion or complete conservation. Increased demands on groundwater have strained the supply in certain regions, resulting in reduced artesian pressure, land subsidence, reduced spring and stream flow, and the intrusion of salt water. This kind of groundwater overdraft is occurring in the High Plains from Texas to Nebraska, parts of California, Arizona, Louisiana, Florida, New Mexico, Arkansas, Wisconsin, Illinois, and North Carolina. It has been reported that groundwater levels of the Carrizo Aquifer in south-central Texas have declined as much as

Table: *Allocation of Water Use In U.S. Communities.*

Use	*Percentage*
Residential	40
Commercial	15
Industrial	25
Public	5
Unaccounted for	15
Total	100

Source: J. M. Last, in Maxcy-Rosenau, eds. *Public Health and Preventive Medicine,*11th ed. (Norwalk, Conn.: Appleton-Century-Crofts, 1980), p. 976. Reprinted by permission of the publisher.

Table: *Allocation of Interior Residential Water Use*

Use	*Percentage*
Drinking and cooking	5
Bathing	30
Toilet flushing	40
Laundry	15
Dishwashing	5
Miscellaneous	5
Total	100

Source: J. M. Last, in Maxcy-Rosenau, eds. *Public Health and Preventive Medicine,*11th ed. Reprinted by permission of the publisher.

400 feet since 1930 because of overpumping (Texas Department of Water Resources, 1984b). In the Second National Water Assessment, the U.S. Water Resources Council (1978a) lists actions that could be taken before pumping becomes uneconomical in the areas with declining water levels These include locating alternate water sources, developing artificial recharge of the aquifers, relocation of water-requiring activities, and reduction of water use through better water management. Groundwater overdraft is considered to be serious when it continues indefinitely, causing the exhaustion of the resource, associated economic and social dislocations, and perhaps general weakening of a region's economy. Such potential problems are developing in California, the High Plains regions, Florida, New Mexico, Colorado, and Arizona, but many states are successfully dealing with the problem (U.S. GAO, 1980). Although this book deals with changes in groundwater quality, changes in quantity through overdraft are briefly mentioned here, as they can affect the quality of the remaining groundwater in the region.

Profile of an Aquifer

In order to provide a picture of the importance of groundwater, it is useful to describe a specific aquifer, its characteristics and uses, and the problems that affect it as a natural resource. A study by the High Plains Study Council provides such data for the Ogallala Aquifer. The Ogallala is an exceptional aquifer in both extent and volume, running as it does through parts of eight states. It is an integral part of the burgeoning economy of those states, as well as of the environment. Because of its unusual size and certain other characteristics, however, it should not be viewed as a representative aquifer.

The Ogallala Aquifer is a major source of groundwater in the High Plains region, underlying a land area of approximately 33 134,000 square miles, three times the size of New York State, or about the size of California. The formation extends from southern South Dakota to northwestern Texas and transects portions of six other states: Nebraska, Wyoming, Colorado, Kansas, Oklahoma, and New Mexico. It contains an estimated quadrillion gallons, or 3.25 billion acre-feet of water, the equivalent of Lake Huron. The aquifer fuels a $30 billion a year agricultural economy; irrigates over 12 million acres of farmland, or 20% of the irrigated acreage in the United States; and helps support a population of nearly half a million people. Furthermore, 40% of the beef consumed in this country is fattened in the aquifer area. Such heavy exploitation of the vast groundwater resources in the Ogallala has resulted in serious depletion in some areas. These problems are now the subject of intensive study.

The Ogallala was formed during the early Pleistocene time, some two million years ago, from glacial outwash of the ancestral Rocky Mountains that consisted of gravel, sand, and finer debris that was caught up by streams of meltwater running away from the glacier. The outwash settled unevenly throughout the High Plains, giving the aquifer water-storage capacities that varied according to the depth and content of the outwash. In Nebraska, where two-thirds of the Ogallala's waters lie and where the aquifer is 1,000-1,500 feet thick, the aquifer has a considerably greater storage capacity per unit of area than it does in the South Plains of Texas, where it is less than 100 feet thick. At present, there are an estimated 2 billion acre-feet of water in storage in Nebraska compared with 350 million acre-feet in Texas and 100 million acre-feet in Oklahoma. Annual withdrawals from the entire aquifer average 5 or 6 million acre-feet (U.S. Water Resources Council, 1980). The Ogallala's extensive groundwater resources were virtually unknown until the early part of this century. The first wells were drilled into the formation over 90 years ago, but the land was used primarily for cattle grazing and dryland wheat production well into the 1900s. The start of irrigated farming, with the use of high-capacity pumps, 34 after World War II brought rapid economic growth but also caused the water table of the aquifer to decline steadily, in some areas by as much as 100 feet, and even to the point of drying up.

Gaines County, Texas, has dried up its groundwater resources entirely. The aquifer is very slow to recharge. Because of its high rate of evaporation and its high percentage of impermeable soils, the water is not replenished at the rate at which it is used. The depth of the water table nearly everywhere in the Ogallala formation is 50 feet or more below the roots of plants. Plants get the first opportunity at capturing any infiltrating rainfall, making that portion unavailable for pumping. Overall, the amount of water being overdrawn annually from the aquifer exceeds 3 million acrefeet. Annual recharge from precipitation averaging 12 to 22 inches per year, irrigation return-flows, and some stream-bed percolation is estimated to be 0.27 million acre-feet. The state of Texas has estimated that, at the current rate of use and without an imported water supply, approximately 40% of the now-irrigated acreage in the High Plains of West Texas will have to revert to dryland farming or be abandoned by the year 2000, and 60% will have to be abandoned by the year 2020. The depletion of the Ogallala began to attract public attention in the early 1970s. Members of Congress representing the Great Plains states of Colorado, Kansas, Nebraska, New Mexico, Oklahoma, and Texas encouraged the development of legislation that would mandate intensive study of the problem. A $6 million study was authorised by public law in 1976, and the High Plains Study Council was formed under contract with Camp, Dresser & McKee. The Council approved a study design with the following objectives in 1977:

1. to determine the potential development alternatives for the High Plains,
2. to identify and describe policies and actions required to carry out promising development strategies, and
3. to evaluate the local, state, and national implications of these alternative development strategies.

It is a comprehensive resource and economic-development study of the area served by the Ogallala. Major emphasis is being given to improving the water supply by local conservation and by improved practices of irrigation and agricultural management and to interbasin transfers of surplus water from adjacent basins, the Missouri River, and streams in Arkansas. Effective, comprehensive, interstate management of the 35 Ogallala groundwater resource is a further major element of the study.

Currently there is little or no public support for comprehensive, interstate management of the aquifer, although interdependencies are widely recognised. Laws in the states vary widely. Oklahoma and Texas are the only states of the High Plains with private ownership of groundwater. The groundwater in the six other states is dedicated to the people of the state, and allocation is administered by state and local officials. In all eight states, well registration or permitting systems exist.

Contamination of the Ogallala aquifer does not, to date, appear to be a major problem or concern. There has been some contamination around major population centres like Lubbock, Texas, but the depletion problem is of more immediate concern. A study by Camp, Dresser & McKee predicts that 5.1 million acres of irrigated land in six Great Plains states will dry up by the year 2020. A continuation of present trends will mean the loss of 1.6 million irrigated acres in Kansas, 1.2 million in Texas, 260,000 in Colorado, 224,000 in New Mexico, and 330,000 in Oklahoma. Twenty recommendations have been made for remedial action, but only one is an actual cure. The Army Corps of Engineers has proposed a system of huge canals that would import water from South Dakota, Missouri, and Arkansas. The cost of the project (from $3.6 billion to $22.6 billion) is prohibitive, however, and stopgap efforts like conversion to dryland farming will probably be the most immediate solutions to the depletion problem.

Uses of the Ogallala Groundwater Resources In a Few Selected States

In New Mexico the aquifer is heavily developed from southern Quay County to southern Lea County. The thickest and most productive part of the aquifer is in northern Lea County. Individual well yields in this area can be as much as 1,600 gallons per minute. In southern Lea County, where the aquifer is thinner, wells yield from 300 to 1,000 gpm, but water levels have been declining at a rate of almost 3 feet per year since 1950. In Oklahoma the Ogallala is the most important aquifer, underlying most of the panhandle. Deposits of sand, gravel, and minor amounts of clay store high-quality 36 groundwater. The aquifer is as much as several hundred feet thick, contains more than 100 million acre-feet of available water, and supplies most of the water requirements of the panhandle. It is used to irrigate 135,000 acres, to supply water for the industrial needs of the Keyes helium plant and the natural gas industry in the panhandle, and to supply all the public and domestic needs in the area. In Texas the aquifer formation consists of interfingered and intergraded lenses and layers of sand, gravel, silt, clay and caliche, a crust of calcium carbonate. The High Plains is divided by the Canadian River into the northern High Plains and the southern High Plains. The northern area comprises approximately 9,300 square miles. The zone of saturation in most places is 100 to 500 feet thick.

Irrigation has developed more slowly here than in the southern High Plains, which is the area of greatest groundwater development in Texas. This area includes about 25,000 square miles and has the most serious depletion problems of the entire aquifer. According to a 1985 USGS report, groundwater withdrawals in the southern High Plains have been averaging 6,500 million gallons a day (mgd) (USGS, 1985). The rate of recharge has been 125 to 134 mgd. The Texas Department of Water Resources has made estimates and projections of groundwater availability in the Ogallala through the year 2030. It has determined the annual effective recharge to be 143 billion gallons.

The amount of water in recoverable storage as of 1980 was determined to be 125,609.27 billion gallons. (Recoverable storage is defined as that portion of the underground reservoir capacity that, it is estimated, is physically capable of yielding water economically.) The projected annual average of groundwater availability was then determined through 2030. (Annual average of groundwater availability is defined as the estimated sustainable annual yield, or effective recharge, plus that amount of water that can be recovered from storage over a specified period of time without causing irreversible harm, such as land-surface subsidence or water-quality deterioration). The projections are as follows:

1990	2,132.17 billion gallons
2000	2,678.33 billion gallons
2010	2,495.95 billion gallons
2020	1,959.99 billion gallons
2030	1,491.09 billion gallons

The remaining recoverable storage in 2031 is projected to be 49,696.48 billion gallons (Texas Department of Water Resources, 1984c).

Groundwater Contamination:

- Groundwater contamination incidents have been reported from all parts of the United States.
- More contamination is likely to be discovered in the future, especially if a comprehensive national survey is undertaken.
- There are many and varied sources of contamination: natural and anthropogenic, point and nonpoint, planned and inadvertent.

- Types of contaminants found in groundwater range from simple inorganic ions, such as chloride, nitrate, and heavy metals, to complex synthetic organic chemicals.
- The types of reported contamination problems, and their sources, vary from one region of the country to another and are influenced by climate, population density, intensity of industrial and agricultural activities, and the hydrogeology of the region.

The mobility of our society, as well as the distribution of industry and agriculture, depends upon an available supply of clean water. Nevertheless, instances of groundwater contamination have been found in most sections of the country. For the purpose of this book groundwater contamination will be defined as the addition to water of elements, compounds, or pathogens that alter its composition.

The pollution of groundwater occurs when the concentrations of the contaminants render the water unfit for present and future uses. One of the major difficulties with groundwater contamination is that it occurs underground, out of sight. Thc sources of pollution are not easily observed, nor are the effect of pollution often seen until irreversible damage has occurred. There are no obvious warning signals such as fish kills, discoloration, or stench that are often early indicators of surface water pollution. Where contamination affects pumping wells, such indicators may occur, although many commonly found contaminants are both colourless and odourless, which makes detection difficult. They occur in concentrations that, if ingested, may cause long-term chronic illness, rather than acute poisoning. Many chemicals now found in groundwater have not been tested for their effects on human health. The tangible effects of groundwater contamination usually come to light long after the incident causing the contamination has occurred. The long time lag between occurrence and detection is a major problem.

Groundwater can be contaminated by a variety of compounds, both natural and man-made. Contamination due to man has occurred for centuries, but industrialisation, urbanisation, and increased population have greatly aggravated the problem in some areas. Miller (1981a), Braids (1981a) and Saar (1985) summarised the mechanisms of contamination and attenuation. Freeze and Cherry (1979) give a detailed account of these mechanisms, as well as of the movement of a contaminant in an aquifer.

Movement of Contaminants

A contaminant usually enters the groundwater system from the surface of the land, percolating down through the aerated soil and nonsaturated zone. The root zone may extend two or three feet into the soil. Many reductive and oxidative biological processes take place in this zone that may degrade or biologically change the contaminants. Plant uptake can remove certain heavy metals; microbial fixation and other biological processes can also remove a fraction of the contaminants, the size of the fraction being dependent on the nature of the contaminant. For example, depending upon conditions, iron and manganese solubility can be affected by microbial oxidation and reduction in the soils. In the deeper geologic material that consists mainly of humus and weathered rocks, there is a lessening of such biological processes. Attenuation of contaminants in this zone may occur by surface adsorption as ions in the contaminants are attracted to the charges on the clay particles. Other contaminants may be removed by insoluble organic matter complexing, giving rise to complexed humic acids. Microbial action may influence redox potentials and cause the release of inorganic ions during decomposition. The susceptibility of different contaminants to differential attenuation varies. Present knowledge indicates that more different attenuation processes are active in the nonsaturated zone than in the aquifer; and that the amounts of degradation are larger. However, many significant bio-conversions are being discovered in anaerobic subsurface environments. In fact many compounds that persist under aerobic conditions are metabolised in anaerobic conditions.

In aquifers composed of unconsolidated granular media, contaminant movement is determined by groundwater flow rates and flow paths which are governed by advection, diffusion, and dispersion, by the density of the contaminant and biological, chemical and physical interactions with the material through which it flows. Cherry (1984) states that "Contaminant migration in fractured media is determined by advection and dispersion in the fractures and by diffusive mass transfer between the flowing water in the fractures and the stagnant water in the porous but low permeable matrix." Specific contaminants move in different ways. Liquid petroleum hydrocarbons, for example, would be expected to move downward through the aquifer; however, they may spread out on the top of the aquifer if they are present in sufficient quantities. They usually do not move very far from the point of entry because of the limiting effect of surface tension between the liquid and the water-coated grain of the unconsolidated material. However, dissolved chemicals from these petroleum hydrocarbons move with the groundwater.

Industrial liquids that are halogenated hydrocarbons have a viscosity less and a density much greater than those of water. The fractions that dissolve in water move with the groundwater. However, they may move downward, settle out in the aquitard, and (depending on the slope) may move downslope and not in the direction of the bulk flow of the groundwater.

The migration of contaminants through unconsolidated granular media is better understood than is their migration through fractured material. Unfortunately, contaminant migration through subsurface materials is a highly complex and poorly understood science. Much of the quantitative data obtained for the prediction of contaminant migration from theoretical models and laboratory data is of questionable use when applied to the field. Problems arise mainly from variable geologic conditions (heterogeneous, anisotropic, and fractured subsurface materials), and from the many types and characteristics of the contaminants themselves. Prediction of plume movement and behaviour is made difficult because geologic formations are not often uniform. Layered beds and lenses may cause fingering or separation of the plume. Different geologic materials may retard or enhance the movement of the plume. The nature of the plume is further affected by the reactivity of the constituents of the contaminant.

Concentrations can be altered by a variety of chemical and biochemical reactions, including adsorption-desorption reactions, acid-base reactions, solution-precipitation reactions, oxidation reduction reactions, ion pairing, and microbial cell synthesis. The plume continues to move with groundwater flow unless it is blocked and eventually reaches points of groundwater discharge such as streams, wetlands, lakes, and tidal waters. Because very little dilution takes place, concentrations of contaminants are often much higher in groundwater than in surface water.

Sources of Contamination

Sources of groundwater contamination may be divided into three main categories:

1. natural pollution
2. waste-disposal practices
3. nondisposal sources due to human activities.

The degree of threat posed by these sources of contamination depends upon the concentration of the contaminant, its toxicity at that concentration, the volume of groundwater affected, the hydrogeological conditions, the uses made of the water from that particular aquifer, the population affected by such uses, and the availability of an alternate water supply.

Groundwater-pollution problems and their sources have been the object of numerous summaries. An incomplete but illustrative listing includes Keeley, 1976; Fuhriman and Barton, 1971; van der van der et al., 1975; Miller et al., 1977; Scalf et al., 1973; Miller et al., 1974; U.S. EPA, 1980a, 1978a, 1978b; U.S. GAO, 1978, 1980; U.S. Water Resources Council, 1978a, 1978b, 1978c. A comprehensive survey of sources of contamination due to wastedisposal practices was completed for the 1977 Report to Congress. However, the most comprehensive listing of sources thus far is that put together by the Office of Technology Assessment. OTA identified 33 sources of known groundwater contaminants and categorised them according to the nature of their release.

Changes in the Composition of Groundwater Due to Natural Processes

All groundwater contains some dissolved salts. Mineralisation of groundwater due to leaching is a significant source of saline groundwater in the arid areas of the country and is greatest in the areas of lowest precipitation. In areas where the water table is near the surface, evaporation and transpiration further concentrate the salts in the remaining water. Natural leaching has been a significant source of contamination in the arid southwest and south-central area comprising Arizona, California, Nevada, Utah, Arkansas, Louisiana, New Mexico, Oklahoma, and Texas. In this area, natural leaching has exceeded in importance anthropogenic sources of groundwater contamination. Many of the inorganic substances found in groundwater are largely the result of natural contamination, although human sources contribute significant amounts as well. Chlorides are a widespread contaminant. Sulfates, nitrates, fluorides, and iron are also common natural contaminants that occur in localised natural deposits, often causing the groundwater in those areas to exceed EPA standards. Radioactivity from uranium deposits has caused problems in Texas, Illinois, Iowa, Oklahoma, Nevada, Arkansas, and New Mexico. Arsenic may be a local problem in the thermal springs of the northwest region (van der Leeden et al., 1975). Other examples of such localised problems occur throughout the country. Water from fault zones or of volcanic origin can contain high levels of salts or toxic chemicals. As mentioned in the section characterising groundwater, the aquitards that confine the aquifers are not totally impermeable. A change in the composition, thickness, or continuity of the aquitard may permit leakage from one aquifer to another. The leaking of a saline aquifer into a potable aquifer may cause deterioration of the quality of water in the potable aquifer. Although natural alteration of the chemical composition of groundwater cannot be prevented, it may be fit for some uses with or without pretreatment.

Contamination Due to Waste-Disposal Practices

The 1984 OTA report to Congress *Protecting the Natio's Groundwater from Contamination* had the best estimate of the number of sources of groundwater contamination from waste disposal at the time that this report was compiled. The major sources described in this report and elsewhere will be summarised below. It should be noted that many of the numbers given are estimates and that definitive data on waste-disposal practices often were not, and are not, readily available. Waste from natural and manufactured products is often stored on or beneath the land surface. In fact, burying wastes seems an ingrained cultural phenomenon. Such burial of wastes is now regulated by the statutes. Sources of contamination are derived from all aspects of our lives, including industry, agriculture, and government. One estimate is that between 50,000 and 75,000 chemicals are now being used and distributed through the environment and that an additional 700 to 800 are being added each year.

Table: *Inorganic Substances Found in Groundwater*

Substance	***Concentration (mg/l)***
Aluminum	0.1-1,200
Ammonia	1.0-900
Antimony	—
Arsenic	0.01-2,100
Barium	2.8-3.8
Beryllium	less than 0.01
Boron	—
Cadmium	0.01-180
Calcium	0.5-225
Chlorides	1.0-49,500
Chromium	0.06-2,740
Cobalt	0.01-0.18
Copper	0.01-2.8
Cyanides	1.05-14
Fluorides	0.12-250
Iron	0.04-6,200
Lead	0.01-5.6
Lithium	—
Magnesium	0.2-70
Manganese	0.1-110
Mercury	0.003-0.01
Molybdenum	0.4-40
Nickel	0,05-0.5
Nitrates	1.42-433
Nitrites	—
Palladium	—
Potassium	0.5-2.4
Phosphates	0.4-33
Selenium	0.6-20
Silver	9.0-330
Sodium	3.1-211
Sulfates	0.2-32,318
Sulfites	—
Thallium	—
Titanium	—
Vanadium	243.0
Zinc	0.1-240

— indicates that the substance has been detected in groundwater but no concentrations have been reported

Source: OTA, 1984.

In 1984, the OTA listed more than 200 substances that have been detected in groundwater; these include 175 organic chemicals and 50 inorganic chemicals, organisms, and radionuclides. For disposal, however, most are reacted with other chemicals, immobilised, or otherwise treated and burned.

Individual Sewage Disposal Systems

The 1980 census estimated that there were approximately 22 million septic systems serving nearly one-third of the nation's population. Annual flow to an individual septic tank ranges from approximately 49,000 to 75,000 gallons per household. The total annual flow to all domestic systems can be estimated to be from 1.08 trillion to 1.65 trillion gallons, making these systems the highest-ranking source of wastewater discharged directly into groundwater. The three methods of onsite domestic waste disposal are septic tanks, with their subsurface disposal system; the less satisfactory cesspool, commonly found in older installations where there is a deep layer of permeable soil; and finally the pit privy.

Properly constructed septic-tank systems and privies permit effective treatment of human waste. Cesspools work only in coarse or highly fissured materials and have a high potential for contamination. The septic tank has several important functions. It separates solid and liquid wastes, stores solids and floatable materials, and treats aerobically both stored solids and nonsettlable material.

Soils with slow percolation rates, shallow soils, soils over permeable bedrock, and soils with permanent or periodic high water tables are not suitable for the use of conventional septic tank systems. The leach beds for septic tanks typically consist of 500 to 600 square feet of drain bed, located 18 to 24 inches below the surface of the ground. The effluent is thus distributed over a wide area, and the shallow depth of the bed permits some evaporation and some uptake of contaminants by plants.

Because it is near the surface, the drain field operates under aerobic conditions, with the result that good quality water reaches the groundwater. It is important that a zone of unsaturated soil occur between the leach bed and the water table so that the effluent from the septic tank is not discharged directly into the groundwater and hence into the aquifer. The degree of potential risk posed by these systems depends in large measure upon their design and installation and upon the hydrogeology of the area. When sited correctly they operate efficiently, renovating wastewater and returning good quality water to the watershed. When operating incorrectly, they can allow pollutants to enter the groundwater. This would be particularly 62 troublesome in areas where drinking water wells tap the same aquifer in the vicinity of the leach bed. The advantages of septic tanks include the following:

- Minimal maintenance is required.
- The cost of individual or community septic tanks is less than the cost of central wastewater collection and treatment plants.
- It is a low-technology system.
- The energy requirements are low when compared with centralised wastewater-treatment facilities.

The disadvantages include:

- The potential for groundwater contamination exists when the tanks are incorrectly sited, with regard to density of the individual units or soil characteristics or both.
- The systems require proper maintenance. Failure to provide this results in system overflows or pollution of nearby wells.
- Septic-tank cleaners may severely contaminate groundwater.

Table: *Characteristics of Domestic Sewage*

Constituent	***Concentration *Typical Domestic Sewage***
* mg/l except for conductivity (micro-mhos/cm), and pH (pH units).	
Total Suspended Solids	200
Conductivity	700
Chemical Oxygen Demand (COD)	500
Biochemical Oxygen Demand (5 day BOD)	200
Total Organic Carbon (TOC)	200
pH	8.0
Alkalinity (as CaCO)	100
Acidity (as CaCO)	20
Total Phosphorus	10
Total Nitrogen	40
Chloride	50
Calcium	50
Magnesium	30
Iron	0.1
Manganese	0.1

Source: U.S. EPA, 1973.

The constituents that pose the greatest threat to the quality of groundwater are nitrates, phosphates, heavy metals, inorganic ions (Na^+, Cl^-, $SO_4^=$, K^+, Ca^{++}, and Mg^{++}), and pathogenic organisms. In addition, toxic synthetic organic chemicals are becoming a more significant hazard due to the increased use over the last ten years of cleaning products containing such chemicals.

It is difficult to show by mapping how regions are affected by sources of contamination. All contamination incidents are indicated by points on a map, but not all the groundwater in the area indicated is contaminated. Since the most important factors affecting the contamination of groundwater by sewage are the density of the individual units and the soil characteristics, the greatest potential for contamination is in the eastern region of the country. A septic-tank density of greater than 40 per square mile constitutes a region of potential contamination to groundwater.

Land Disposal of Solid Wastes

According to the Resource Conservation and Recovery Act of 1976, solid wastes are defined as any garbage, refuse, or sludge from waste-treatment plants, water-supply treatment plants, or airpollution control facilities, and any other discarded materials, including solid, liquid, semi-solid, or contained gaseous materials, resulting from industrial, commercial, mining, and agricultural operations and from community activities.

They do not include solid or dissolved material from domestic sewage, irrigation returnflows or industrial discharges that are point sources, or products and by-products of the nuclear industry.

Solid wastes are considered hazardous when their quantity, concentration, or physical, chemical, or infectious characteristics cause, or significantly contribute to, an increase in mortality or an increase in serious illness, or pose present or potential hazards to human health or the environment when improperly treated, stored, transported, disposed of, or otherwise managed.

The solid portion of household wastes contains a high proportion of putrescible matter that is broken down by biodegradation, releasing carbon dioxide and methane gas (UNESCO, 1980). The leachate contains high concentrations of sulfate, chloride, and ammonia. Abundant quantities of cellulose from paper products may retard the movement of halogenated hydrocarbons by absorption.

Solid commercial wastes have a similar composition to household wastes. They may, however, contain greater quantities of 64 oils, phenols, and hydrocarbon solvents. Phenols are the most resistant to biological breakdown and may be leached. Domestic and commercial wastes are serious groundwater contaminants because of the numerous dissolved constituents and the high biological oxygen demand (BOD) of may of the constituents. The table does not include synthetic organic chemicals.

***Figure** : Solid Waste Disposal Land*

The solid components of industrial wastes vary with the source of production (UNESCO, 1980). Cyanide wastes are produced in metallurgical operations; sulfite wastes come from paper and pulp manufacturing; mercury is a waste product in the electrical industry; and the petrochemical industry produces several solid residues ranging from polychlorinated biphenyls (PCBs) to pesticides and herbicide residues to phenol-rich tar wastes. The amount of solid waste disposed of by various manufacturing industries is listed.

Other serious solid-waste contaminants can result from thermal power generation from the burning of coal, forming fly ash which, due to its high surface area volume ratio, is fairly reactive.

Leaching of the deposited fly ash may initially give rise to high concentrations of arsenic, chromium, selenium, and chloride. Another contaminating product resulting from the burning of coal is the sludge formed by the aqueous scrubbing of flue gases. Sludges typically contain cyanide and heavy metals and are of low pH unless neutralised by the addition of lime. It has been found that mixtures of sludge, fly ash, and lime rapidly set into a low permeability solid which leaches less readily. Contamination occurs when precipitation infiltrates and percolates through solid wastes at poorly designed land-disposal sites, forming a leachate of dissolved minerals, heavy metals, and organic chemicals. Leachate formed at disposal sites which are located in wetlands, flood plains, or where there is a shallow water table and that are constructed without natural or artificial barriers is likely to contaminate groundwater.

The types of land disposal sites for solid wastes include dumps, landfills, sanitary landfills, and secured landfills listed in increasing order of ability to protect the environment from the adverse effects of their use. Landfill regulations promulgated under the Resource Conservation and Recovery Act (RCRA) were issued in July 1982. The Comprehensive Environmental Response, Compensation and Liability Act (CERCLA) requires remedial actions at inactive sites that pose a threat to the environment and to human health. The U.S. EPA estimates that there are 93,000 landfills in the United States. Approximately 75,000 of these are onsite industrial landfills and 18,500 receive municipal wastes (U.S. EPA, 1985c). In 1983, another survey estimated that there are between 15,000 and 20,000 active landfills in the nation, including municipal landfills, open dumps, and nonhazardous waste facilities, operating, abandoned, or closed. (OTA, 1984).

***Table:** Summary of Leachate Characteristics Based on 20 Samples from Municipal Solid Wastes*

Components	*Median Value (ppm)*[a]	*Ranges of All Values (ppm)*[a]	
Alkalinity ($CaCO_3$)	3,050	0	-20,850
Biochemical Oxygen Demand(5-day BOD)	5,700	81	-33,360
Calcium (Ca)	438	60	- 7,200
Chemical Oxygen Demand (COD)	8,100	40	-89,520
Copper (Cu)	0.5	0	- 9.9
Chloride (Cl)	700	4.7	- 2,500
Hardness ($CaCO_3$)	2,750	0	-22,800
Iron, Total (Fe)	84	0	- 2,820
Lead (Pb)	0.75	<0.1	- 2.0
Magnesium (Mg)	230	17	-15,600
Manganese (Mn)	0.22	0	- 125
Nitrogen (NH_4)	218	0.06	- 1,106
Potassium (K)	371	28	- 3,770
Sodium (Na)	767	0	- 7,700
Sulfate (SO_4)	47	1	- 1,558
Total Dissolved Solids (TDS)	8,955	584	-44,900
Total Suspended Solids (TSS)	220	10	-26,500
Total Phosphate (PO_4)	10.1	0	- 130
Zinc (Zn)	3.5	0	- 370
pH	5.8	3.7	- 8.5

[a] Where applicable

Source: U.S. EPA, 1977.

Hazardous wastes may be disposed of in a landfill, dump, or impoundment. EPA requires that site locations and amounts and types of wastes stored be reported, under CERCLA by owners and operators of inactive sites, and under RCRA by present owners and operators of active privately owned industrial sites. About 199 active hazardous-waste landfill facilities (OTA, 1984) and approximately 20,000 abandoned hazardous-waste sites (EPA, 1985c) are known to exist across the country. The amount of solid waste generated in this country has been estimated to be enough to cover 400 acres of land to a depth of 10 feet each day. It is estimated that in 1980, 3.74 pounds of solid waste were generated per person per day, amounting to 155.6 million tons of municipal wastes nationwide for the year. OTA (1984) also estimates that 30 million tons of solid wastes generated by utilities and 1 million tons of municipal sludge are disposed of in landfills per year. Between 40 million and 140 million wet tons of nonhazardous industrial solid wastes are disposed of annually (OTA, 1984). In 1981 hazardous-waste facilities received 14.7 billion gallons of both liquid and solid hazardous wastes (CEQ, 1984).

Collection, Treatment, and Disposal of Municipal Wastewater

The collection, treatment, and disposal of municipal wastewater may be a problem in urban areas. Municipal wastewaters include domestic wastewater, storm water and its associated debris, and industrial wastes. They are collected by sanitary sewer systems and transported to treatment sites. Approximately 170 million Americans were served by sewer systems in 1984. Leaks may occur in the system due to design, age, disruption by tree roots, seismic activity, or poor construction. The lagoons and ponds used in wastewater treatment operate under anaerobic, aerobic, or facultative conditions.

Figure : *Using treated municipal wastewater can be quite difficult*

The total volume of sewage handled by one of these plants is estimated to be approximately 0.667 mgd. Lagoons are often constructed without adequate seals, thus promoting leakage under certain geologic conditions and posing a threat to groundwater. Another possible route of groundwater contamination is by the land disposal of treated wastewater, which could alter the chemical constituents of the natural groundwater. Land disposal methods include agricultural irrigation, rapid infiltration ponds, overland runoff, and discharge into dry stream beds and ditches. Methods involving spraying onto land are not suitable for use in freezing winter months.

Industrial and Other Wastewater Impoundments

Under the Safe Drinking Water Act of 1974, EPA undertook a national survey of surface impoundments, which was carried out from 1976 to 1978, and the interim report was published in 1978. The final report, released in 1983, contains data from 1978 through 1980. These reports define surface impoundments as depressions in the land (pits, ponds, lagoons, and pools) containing liquid, semi-solid, and solid wastes. In the final report EPA stated that the number of sites located in the United States totalled approximately 80,263 and contained more than 180,000 impoundments.

Thirty-one percent of the sites were oil 68 and gas brine pits, 25% municipal sites, 18% agricultural sites, 15% industrial sites, and 9% mining sites. At the time of the survey it was estimated that approximately 70% of the industrial sites, 84% of the agricultural sites, and 78% of the municipal sites remained unlined.

More than 98% of the sites were located in areas with proximity to potential water supplies. Approximately 15% of the sites (excluding oil- and gas-related facilities) contained hazardous wastes. New Mexico was estimated to have the largest number of sites, with 12,901, and Rhode Island the fewest, with 44. The regulations under RCRA may have affected the validity of these estimates, but no recent figures are available.

Table: *Summary Statistics for Located Active Surface Impoundment Sites*

Category	*Located Sites**		*Assessed Sites**		*Located Impoundments**	
Industrial	11,760	(15%)	8,662	(28%)	27,912	(15%)
Municipal	19,746	(25%)	10,822	(34%)	37,185	(21%)
Agricultural	14,850	(18%)	6,646	(21%)	19,437	(11%)
Mining	7,364	(9%)	1,552	(5%)	25,038	(14%)
Oil & Gas Brine Pits	24,990*	(31%)	3,354	(11%)	65,488	(36%)
Other	1,553	(2%)	350	(1%)	5,913	(3%)
Total	80,263		31,386		180,973	

* SIA site numbers for the mining and oil & gas brine pit sites are usually related to lease or field data, not to actual ownership, and should not be referred to as the actual number of legal sites. The number of located impoundments would be a closer approximation for these two categories. *Located sites:* Total number of facilities identified in the inventory. *Assessed sites:* Total number of facilities evaluated in the inventory. *Located impoundments:* Total number of impoundments identified in the inventory. The number is larger than located sites since many facilities had more than one impoundment.

Source: U.S. EPA, 1983.

***Table:** Estimated Number of Located Impoundments and Impoundment Sites for All Categories, by State*

State	*Active Sites*	*Impoundments*	*Abandoned Sites*	*Impoundments*
Alabama	1,063	1,696		
Alaska	129	179	27	31
Arizona	641	1,615	20	48
Arkansas	861	6,806	12	24
California	1,750	7,577	31	101
Colorado	1,267	3,631		
Connecticut	305	758	83	188
Delaware	52	131		
Florida	3,644	5,610		
Georgia	1,363	1,640	65	70
Hawaii	94	296		
Idaho	253	538		
Illinois	4,471	6,677	591	745
Indiana	1,897	2,678		
Iowa	1,784	2,539		
Kansas	1,970	3,328		
Kentucky	640	922		
Louisiana	984	2,804	49	77
Maine	132	319	41	133
Maryland	443	836	18	25
Massachusetts	341	2,211	38	208
Michigan	2,187	3,505	76	135
Minnesota	1,952	2,733	135	160
Mississippi	1,306	1,667	52	58
Missouri	3,657	4,683	507	507
Montana	668	1,266		
Nebraska	1,632	3,254		
Nevada	216	541	10	382
New Hampshire	152	274	2	4
New Jersey	311	896	46	129
New Mexico	12,901	17,746	39	116
New York	1,070	1,837	40	52

North Carolina	709	1,051	26	30
North Dakota	667	1,168		
Ohio	3,287	16,537		
Oklahoma	2,935	4,538		
Oregon	370	714		
Pennsylvania	7,346	34,224	234	627
Rhode Island	44	133	3	13
South Carolina	1,868	2,259		
South Dakota	606	926		

Table: *Estimated Number of Located Impoundments and Impoundment Sites for All Categories, by State* (continued)

State	***Active Sites***	***Impoundments***	***Abandoned Sites***	***Impoundments***
Tennessee	746	1,325	59	151
Texas	3,672	10,740	170	365
Utah	211	2,345		
Vermont	214	274	10	17
Virginia	1,212	1,975	30	62
Washington	403	1,049		
West Virginia	688	1,861	66	186
Wisconsin	1,022	1,720	49	82
Wyoming	1,146	1,738		
Total	79,818	180,328	2,519	4,726

Source: U.S. EPA, 1983.

Impoundments which are used for the treatment, storage, or disposal of wastes are natural or man-made, and may or may not be lined. Discharging impoundments are designed to discharge regularly into bays, oceans, lakes, or streams. The liquid in nondischarging impoundments is lost by evaporation or by seepage into the soil. The types of wastes received by these sites range from sewage wastes, industrial wastes including those from air scrubbers, cooling tower blow down, and ash residues and oiland gas-extraction wastes to animal feedlot and other agricultural wastes. Of these, the largest users of surface impoundments are the oil and gas industries. Evaporation ponds were used extensively in the South for oil-brine disposal. Evaporation ponds usually lose more fluid by infiltration than by evaporation, especially in the more humid regions of the country. EPA promulgated regulations for hazardous waste surface impoundments in July 1982. In addition, state regulations exist in most oil-producing states.

The chemical character of the waste varies enormously and may include suspended and dissolved solids, pathogenic organisms, oil and grease, detergents, heavy metals, and toxic organic chemicals. It may have a high biological oxygen demand (BOD), chemical oxygen demand (COD), and total organic carbon (TOC). The pH also varies. Not all surface impoundments contain hazardous wastes. Some are used as holding ponds prior to the treatment of the fluids. Because the chemical character of the wastes varies, only a minimal number of impoundments would hold all of the contaminants listed. The type of groundwater contamination known to occur as a result of leaking impoundments reflects the nature of the fluids in the impoundments. Dissolved materials, both inorganic and organic, move into groundwater by direct seepage, while solids may be leached by precipitation or waste fluids. As with all sources of contamination, the sorptive capacity and low permeability of some soils may slow or impede the movement of some of the contaminants.

Land Spreading of Sludge

When the wastewater from municipal and industrial wastes is treated, a residue of sludge remains. Beginning in November 1981, all facilities disposing of sludge considered hazardous on land were required to have a groundwater-monitoring system in place under Resource Conservation and Recovery Act regulations. In 1982 there were at least 2,463 publicly owned treatment facilities applying liquid and thickened sludge on land surfaces, and an additional 485 facilities in operation or under construction using sludge spray irrigation practices (OTA, 1984). Of the 6.8 million dry tons of sludge produced by municipalities in 1982, 24%-29% was spread directly onto crops (OTA 1984). The purpose of land treatments is to biodegrade the organics and to immobilise the inorganics.

The potential for contamination exists, however, since the chemical constituents and viruses in sludge may be leached by precipitation after the sludge is spread on the land. Sludge from manufacturing industries may contain either useful agricultural chemicals or toxic compounds, and it varies from degradable to very refractory. Sludge is stabilised and may be dewatered before land application. The higher the proportion of clay and organic colloids in the soil, the greater the capacity for the heavy-metal ions present in sludge to be immobilised. The main hazardous constituents of the sludges include heavy metals and organic chemicals, such as dyes, inks, oils, pesticides, detergents, organic solvents, and polynuclear aromatic hydrocarbons (PAHs); organic chemicals are typically trace contaminants.

Sludges from the municipal treatment of wastewater or the biotreatment of industrial wastewater include biological components such as bacteria, viruses, fungi, algae, protozoans, rotifers, and other parasites such as worms and flukes. The constituents that may be found in various types of industrial wastes are enumerated in the 1977 Report to Congress. As of 1985, EPA reported seven states that had experienced groundwater contamination due to land spreading of sludges.

Land farming of sludge has been used for nearly 30 years by the petroleum industry. From its research on this method of disposing of oil sludge, the Solid Waste Management Committee of the American Petroleum Institute concludes that if the area farmed is carefully matched to the amount of organic material that must be degraded, the type of land farming practiced, and the degradation rate of the sludge, then the underlying soil and groundwater will not be significantly contaminated.

Brine Disposal Associated with the Petroleum Industry

Groundwater contamination associated with oil production has been documented in at least 21 states (OTA, 1984). During 1983 the total production of crude petroleum was more than 3 billion barrels, valued at more than $83 billion. The principal oil-producing states in 1983 were Texas, Arkansas, Louisiana, and California. Brief mention of the disposal of brines by the use of surface impoundments was made above. Oil production is usually accompanied by the production of saline wastewater in amounts that vary with production procedures. One estimate is that an old well may produce 100 barrels of brine for each barrel of oil. Others estimate that the production of 8 million barrels of crude oil produces 30 million barrels of saltwater and that the ratio of brine to crude oil recovered is 10:1. Still others estimate a ratio of 4:1 brine to oil (OTA, 1984).

In the early days of oil and gas production, the brine was often disposed of in unlined pits. The extensive use of such "evaporation" pits that leaked their contents has caused groundwater contamination in the oil-producing regions. Use of unlined pits has been completely eliminated in some states and partially eliminated in others. For example, unlined pits were banned in Texas in 1969. Other states permit the use of small pits in remote areas if they pose no threat to groundwater. Present practices for brine disposal include injection of the brine into deep underground formations that are deemed unsuitable for other purposes and reinjection of brines into oil-producing formations to enhance oil recovery. These injection practices, if not properly controlled, pose a potential threat to groundwater.

Past oil production practices, including well drilling, completion, and abandonment practices, also were not as strictly controlled as they are now. Some wells were improperly equipped, and abandoned wells were improperly plugged, thereby providing a potential for groundwater contamination. In 1983 abandoned production wells numbered around 1.2 million (OTA, 1984). The need to protect groundwater was recognised, however, and state oil and gas regulatory agencies now have regulations concerning brine injection, well completion, and plugging practices. The Interstate Oil Commission sponsored a study in 1978 that concluded that in Texas, Arkansas, Louisiana, New Mexico, and Oklahoma, reinjection of brines had resulted in fewer instances of groundwater contamination than had the use of evaporation pits. It therefore believes that the practice does not pose a major threat to the groundwater supplies.

Most oilfield brines are corrosive, but the corrosion rate of old and new well casings is unpredictable, and frequent monitoring and surveillance is therefore carried out. Although brine is the main threat to the quality of groundwater related to oil production, minor threats are also posed by the oil and gas themselves, by drilling fluids, by chemicals used in treating wells, by corrosion inhibitors, and by other chemicals. Oil drilling may entail drilling through freshwater aquifers, thus raising the potential for contamination.

Disposal of Mine Wastes

Mining creates conditions and products conducive to groundwater contamination. The recharge and movement of groundwater may be affected by the mining process. Domestic mining for minerals other than organic fuels was more than a $20 billion industry in 1983. Domestic mining for minerals other than organic fuels was a $10 billion industry in 1974. Coal has enjoyed a comeback as an energy source due to the goal of increased self-sufficiency in energy, and it is estimated that between one-third and one-half of the world's total reserves are found in the United States. Techniques of surface and underground mining pose different threats to groundwater quality, although the threat from waste disposal from the two is similar. Although such mining changes the environment, good reclamation may stabilise some of these changes without causing serious damage to the groundwater. Sand, clay, gravel, and stone quarrying are not considered a serious threat to groundwater quality. Threats from the remaining categories of mines include those posed by leachate from tailings piles and drainage from strip, surface, and underground mines. Dewatering of mines, involving a regional lowering of the water table, may have three results: a decrease in the amount of groundwater supplied, a lowering of the water level below the intakes of productive wells, and the oxidation of exposed minerals. Coal is often found in conjunction with iron pyrites (FeS_2) and other sulfides, and the oxidation of these sulfides and the presence of percolating waters produce sulfuric acid and an increase in Fe^{++} and $SO_4^{=}$, leading to acidic groundwater, a serious problem in the coal-mining areas of Pennsylvania and Appalachia. It is very costly either to neutralise acid mine drainage or to backfill abandoned mines in order to prevent the problem. Sealing the mines with airor watertight seals has not proved very successful. A report by the National Academy of Sciences summarizes these and other effects that coal mining may have on groundwater quality.

Metal mining produces drainage waters that have lower concentrations of iron and sulfate than do colliery waters, but that have higher concentrations of dissolved heavy metals due to the lower pH caused by the oxidation of metal sulfides. Wastewater from metal mining may also contain organic flocculants from the onsite processes of screening and dressing the ores. The wastes from such operations, which can be both toxic and radioactive, may contaminate groundwater. The tailings ponds used in the waste-disposal operations of mines may contribute contaminated leachate.

The waste products of quarrying for stone, lime, gypsum, and the like are generally inert, and their only contributions to the water that percolates through them is the possibility of an occasional increase in suspended-solids if crushed material is present. Quarrying does, however, increase the potential for local groundwater contamination in other ways. The removal of soil reduces the possibility of an attenuation of any spills or leakages that may occur in the area and increases the potential for contamination, especially if the bedrock is fractured.

Because of the serious nature of surface water contamination by mining activities, more attention has been given to it than to groundwater contamination, although aquifers in regions with a long history of mining have been written off as sources of water supply. Since then coal mining on a large scale has been initiated in the West, particularly in North Dakota, Wyoming, and Montana. The EPA (1977) estimated that 3.6 million tons of acid may be generated annually from the approximately 200,000 acres used across the nation for disposal of coal-mining wastes. The EPA (1977) also estimated that 10% of the acid enters groundwater each year. Mining operations generate approximately 2.3 billion tons of total waste material annually, including radioactive and nonradioactive waste piles and tailings. The most serious threat to groundwater is posed by the metal mines, particularly uranium and copper mines, which produce waste containing dissolved toxic materials (e.g., arsenic, sulfuric acid, copper, selenium, and molybdenum), as well as radioactive materials.

Deep-Well Disposal of Liquid Wastes

The injection of a variety of liquid wastes into deep wells to avoid having them contaminate the biosphere was becoming a widely adopted waste-disposal practice in the late 1970s, mainly because there were stringent regulations governing disposal into surface waters.

Such wells are less widely accepted now in certain states. It has been estimated that there are more than 221,000 wells in the United States injecting liquid wastes underground: approximately 280 are used for hazardous wastes; 140,000 are used either for brine disposal or for the injection of fluids used in enhanced oil-recovery process; 40,000 are used for agricultural drainage, urban runoff and sewage disposal; and at least 12,000 are in operation for solution mining (OTA, 1984).

Because of the hazardous nature of the injected fluid often radioactive, toxic chemical, petrochemical, and pharmaceutical wastes that are difficult to treat deep-well disposal may pose a great threat if improperly practiced. Eighteen states have already experienced contamination through injection wells. The Underground Injection Control Program of the Safe Drinking Water Act, which prohibits direct injection into drinking water aquifers, and regulations issued under the Resource Conservation and Recovery Act provide such controls.

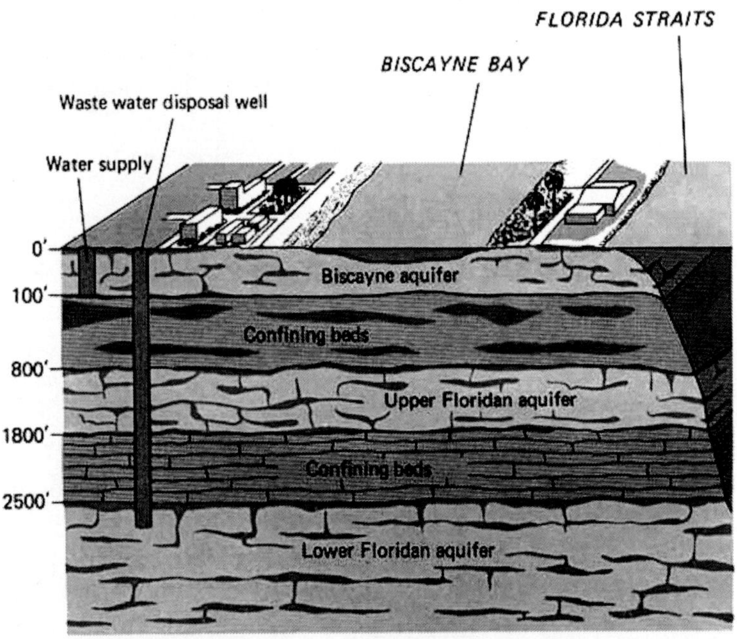

Figure : *Some sewage effluent is disposed of by deep-well injection*

All injection wells are in the depth range of 660 to 13,200 feet, most being between 990 to 6,600 feet. Injection zones are usually located in sandstone, basalt, and carbonate rocks. Injection occurs under pressure, and injection rates vary from 500 to 370 gallons per minute. Injection causes hydrodynamic changes in the aquifer, including the formation of a mound in the potentiometric surface extending in the direction of the regional flow in the aquifer. This is essentially the reverse of what happens when a well is pumped in a confined aquifer. Until 1976, documented cases of disposal-system failure were rare, but they may increase if underground injection becomes more common and occurs over longer periods of time. Underground injection in the vicinity of old unplugged wells, the locations of which are often not known, may lead to the upward leakage of the wastes.

These sources estimate that there may be more than one million unplugged, unlocated wells in North America that were used for purposes other than disposal. Another hazard associated with the practice of waste injection but only indirectly linked to groundwater contamination is the triggering of earthquakes due to increased pore-water pressures along fault lines. The most publicised occurrence, at the Rocky Mountain Arsenal in Colorado, occurred after injection of chemical-warfare waste, using a 12,000 foot well, into fractured pre-cambrian rocks.

The earthquakes thus caused stopped when injection ceased. Other problems may arise if an injection well becomes clogged with injected suspended solids, corrosion products, or precipitation forming between injection fluid and connate water in the aquifer.

Insufficient treatment of injected water may lead to bacterial clogging, and swelling of the mineral constituents of the geologic formation. The Report to Congress stated that injection wells can cause groundwater contamination through the following mechanisms:

- direct injection into drinking-water aquifers
- leakage into potable aquifers due to well-construction faults or failure
- leakage through confining beds by hydraulic fracturing or insufficient thickness
- displacement of saline water into potable aquifers
- migration to potable water zone of the same aquifer.

The report concluded, however, that properly constructed, sited, and monitored waste injection wells can be operated with little danger of groundwater contamination.

Disposal of Wastes from Animal Feedlots

Groundwater contamination by the manure from high-density animal-feeding operations is a relatively new phenomenon. In addition to the threat to groundwater from the vast quantities of manure thus generated, there is also a threat from the food additives, such as hormones and antibiotics, which the manure may contain. As recently as 1977 it was claimed that feedlots pose no danger of groundwater contamination in the southeastern states due to the soil's ability to denitrify the wastes.

Feedlots for poultry and hog farms in Arizona, California, Nevada, and Utah were thought to lead to water-quality problems from bacteria, viruses, nitrate concentrations and with general colour, taste, and odour. The main beef-raising areas, however, are the Corn Belt and the High Plains. Until the 1950s and 1960s, beef was raised on pasture, and the wastes were easily assimilated naturally and posed no problem.

The need for more meat at reasonable cost led to the establishment of feedlots, with capacities of between 1,000 and 50,000 head of cattle. During the four to five months each animal spends in the lot, it produces 0.5 tons of manure (dry weight). This can lead to high nitrate concentrations in the groundwater under some conditions. Where the feedlots are located in areas with a deep water table, the risk of contamination is minimised. The main poultry-rearing region is the Mid-Atlantic and South, in Delaware, Maryland, Virginia, Arkansas, Georgia, and Mississippi, for hogs it is the Midwest, and for sheep the Far West.

Of the potential contaminants in manure, nitrate is the most important, as it is soluble in water and its concentration remains unchanged while passing through the soil. The other contaminants from manure bacteria and phosphate are highly attenuated by the soil and pose less threat. Feedlot management, including stocking rate, density, and manure removal, plays an important role in the protection of groundwater quality.

Heavy manure accumulations may produce an impermeable mat, which in turn produces anaerobic conditions that favour denitrification. Thus nitrate is volatilisied, and little infiltration takes place, minimising the nitrate-contamination potential. Concentrated animal feeding operations are regulated by the Federal Water Pollution Control Act Amendments of 1972 and may require a permit as a point source under the National Pollutant Discharge Elimination System (NPDES).

Contamination from Radioactive Sources

Radionuclides from natural sources, radioactive fallout, nuclear testing and accidents, nuclear fuel cycling, commercial and industrial products and wastes pose a threat to groundwater quality. Almost all groundwater contains some amount of naturally occuring radioactive substances. Types and levels vary geographically and depend on local geology. The highest concentrations of natural radiation of groundwater are often found near granite and phosphorus formations.

Radioactive wastes from the production and generation of nuclear fuel and radioactive materials can be categorised into five types: spent fuel, high level wastes, transuranic wastes, low level wastes and uranium mill tailings. Spent nuclear fuel is irradiated fuel resulting from nuclear power plant operations and is highly radioactive. Most of these wastes are presently being stored in pools at reactor sites until underground repositories are found for permanent disposal. High level wastes result from the initial processing of spent fuel and must be stored carefully in specially constructed facilities. Presently these wastes are being temporarily stored at four utility and defence sites awaiting permanent disposal.

Table: *Current and Projected Quantifies of Radioactive Waste and Spent Fuel (September 1983)*

	Volume of Waste (1000 m)		
Type of Waste	***1983***	***2000***	***2010***
High Level Waste			
Commercial	2.3	.436	3.17
Defence	304	294	269
Transuranic Waste			
Commercial	—	4.6	49.4
Defence	246	340	397
Spent Fuel			
Commercial	4.626	19.378	33.258
Defence	0	0	0
Lower Level Waste			
Commercial	1,020	3,393	5,415
Defence	2,060	3,720	4,710
Inactive Uranium Mill			
Tailings	14,314	23,700	23,700
Active Mill Tailings	96,500	188,800	280,300

(DOE/Defence tailings are about 35% of these totals.)

Source: Adapted from *Spent Fuel and Radioactive Waste Inventories, Projections and Characteristics.* U.S. Department of Energy, 1984.

1. No reprocessing.
2. Found primarily at inactive uranium mills located in western United States. DOE responsible for stabilisation and control of mill tailings in safe and environmentally sound manner under the Uranium Mill Tailings Radiation Control Act of 1978. Anticipated cleanup completion in late 1980s.
3. Located at the 16 active licensed uranium mills operating in 1983, all in the western United States.

Deaf Smith County, Texas, Yucca Mountain, Nevada, or Hanford, Washington. The site is scheduled to be complete by 1998 (Environmental Reporter, 1986). Groundwater contamination by high level radionuclides has been reported in a few instances, largely due to the improper storage of wastes. For example, in the early 1950s, one hundred gallons of high level radioactive wastes leaked from a tank at the Savannah River defence storage site, contaminating some groundwater. In 1973 a 115,000-gallon leak of high level radioactive wastes occurred at the Hanford Nuclear Reservation.

No groundwater contamination resulted, although the soil beneath the site remains radioactive. Another extensive leak of high level radioactive wastes occurred in 1956, spilling 450,000 gallons of high level radioactive wastes in the ground at Hanford, Washington, but with no serious contamination resulting.

Transuranic wastes contain alpha-emitting radionuclides of atomic numbers greater than 92. They result primarily from fuel reprocessing and the manufacture of plutonium containing products (OTA, 1984). These wastes take a long time to decay and require long-term isolation for proper disposal. Presently, there are a total of 7 sites where transuranic wastes may be deposited (OTA, 1984).

Low level wastes exist in all phases and consist of a wide range of waste material. Most of the radionuclides in low level wastes are short lived and have low radioactivity, but are produced in potentially hazardous amounts.

These wastes are generated by hospitals, laboratories, industrial plants, nuclear power plants, and government and defence laboratories and reactors. In the past, low level wastes were either buried in trenches at government sites or dumped out at sea. Now all commercially produced low level wastes are permanently disposed of in one of the three commercial shallow nuclear burial facilities located in Beatty, Nevada, Barnwell, South Carolina, and Hanford, Washington, the major site being in Hanford.

The Department of Energy and the Department of Defence also maintain a number of sites operated by private contractors for low level disposal (OTA, 1984). Shallow burial of low level wastes present a great potential for contamination. Groundwater may erode the structure of the burial site thus providing avenues for contamination to seep. A recent study found that water has come into contact with wastes in burial trenches at 6 of the 11 major low level governmental and commercial burial facilities.

Uranium mill tailings are earthen residues, often in the form of sand, left after uranium is extracted from ores. Disposal commonly takes place in shallow burial grounds located near refineries. Remedial actions at inactive mill tailing sites are to be conducted by the Department of Energy under the Uranium Mill Tailings Radiation Control Act (OTA, 1984).

According to the League of Women Voters (1985) twenty-four abandoned disposal sites are scheduled for clean up by DOE in Arizona, Colorado, New Mexico, North Dakota, Oregon, Pennsylvania, South Dakota, Texas, Utah, Washington, and Wyoming. Of the 28 active uranium mills licensed in the US, 21 reported some degree of groundwater contamination. Remedial action has been taken at 16 of the sites so far.

Other Sources of Contamination Accidental Leaks and Spills

Accidental leaks and spills can occur because of breaks in pipelines, leaks in underground storage tanks, spills from tanker trucks, and other accidents involving the transport and storage of materials. They commonly occur at airports, industrial sites, along highways, railroad sites, gas stations, and refineries.

Small hydrocarbon spills may be absorbed by the unsaturated soil zone, but large ones can reach the water table aquifer and float on the surface. Very small concentrations of petroleum products will render groundwater unfit to drink because of its objectionable odor and taste and also pose a threat to health. Other chemicals have toxic properties. Ammonia, for example, increases the nitrification of groundwater, and acids speed up the solution of heavy metals and soil solids.

The U.S. General Accounting Office (1984) reported that leaks from underground storage tanks (USTs) have been reported in all 50 states; many states claim that this source is their leading cause of underground contamination. There are an estimated 2.5 million underground storage tanks across the U.S., containing more than 25 billion gallons of nonhazardous liquid products, and an additional 2,031 hazardous storage and treatment tanks containing an estimated 13.8 billion gallons of substances (OTA, 1984).

Data was collected at the state level on the numbers of release incidents for EPA's (1986) report *Summary of State Reports on Releases for Underground Storage Tanks*. State files identified 12,444 release incidents that had occurred across the country as of 1984. Even though this data represents the most comprehensive currently available information on leaking USTs, these numbers cannot be regarded as representing all storage tank releases. Differences in state enforcement and reporting procedures have resulted in inherent biases in the data. The results of the study showed that the greatest number of incidents occurred in the Northeast, Mid-Atlantic, and Great Lake states.

This is probably due to the more corrosive conditions of the soils, the higher tank population, and perhaps more efficient procedures for reporting and documenting leaks. Soils became contaminated in 68% of the cases, and groundwater contamination resulted from 45% of the releases. It was found that approximately 65% of the incidents originated from retail gasoline stations, and only 3% of the releases involved chemicals other than petroleum fuels. Ninty-five percent of the leaks were reported to have taken place from operating facilities. The mean tank-leakage age was found to be 17 years for all of the tanks. Eighty-one percent of the tanks were steel and the remaining 19% fibreglass. The most commonly reported cause for release was corrosion and structural failure.

The NAS estimated that 16,000 spills occur annually during transport operations (OTA, 1984). In 1981 alone, 10,072 spills from both oil and hazardous chemicals, totalling 19.6 million gallons were reported (OTA, 1984). The Department of Transportation reported that more than 9,000 spills of hazardous materials occurred during transportation in 1981 and 6,500 in 1982 (OTA, 1984). Emergency measures to deal with spills often consist of flushing the spilled compound away, thus permitting or even assisting in the contamination of groundwater. Flushing usually reduces the immediate threat of fire and explosion.

Such practices are more likely to create surface water pollution than groundwater pollution, as most storm sewers and ditches flow into surface water. If the spill is large enough and is not properly cleaned up, however, groundwater contamination can result.

Under the Clean Water Act, Spill Prevention Countermeasure Control (SPCC) plans have been required of every facility since 1975 for spills to the ground. The National Contingency Plan, under CERCLA, effective December 1982, details cleanup procedures for spills of oil and hazardous materials on land, into surface water, or into groundwater.

In addition, releases of quantities of oil or hazardous material in excess of stated amounts must be made known to the National Response Centre. Many of the industrial trade associations have researched cleanup and containment and have published manuals. For example, the American Petroleum Institute manual for the cleanup of underground spills (API, 1980).

Pipelines are used to transport a variety of fluids, but the most common one is petroleum. In 1972 there were 174,000 miles of petroleum pipelines that transported a total of 8.5 billion barrels of petroleum products. In 1976 the same amount of pipeline transported 9.63 billion barrels of petroleum products. In 1971 there were 308 interstate pipeline accidents, involving a loss of about 245,000 barrels of liquids.

Pipeline failure due to external corrosion caused 33% of these accidents, and damage to pipeline from excavating machines caused an additional 22%. In 1980 there were 275 accidents; in 1981, 198 accidents. The decrease in incidents may be due to extensive regulations from the Department of Transportation regulating transportation of liquids by pipeline.

Of the 4,112 accidents that occurred between 1968 and 1981, 1,372 were due to corrosion and 1,101 to pipeline 83 ruptures. No information was given about the number of barrels involved in the incidents in 1980 and 1981. In some regions deliberate leakage and spillage of contaminants, so-called midnight dumping, has occurred. This source of contamination, which is a criminal offence, is considered to be a major problem in New England.

Agricultural Activities

Irrigation return-flow; the use of pesticides, fertilizers, and manure; and changes in vegetative cover have all been known to cause changes in groundwater quality. Contamination of groundwater by nitrates is one of the largest sources of nonpoint pollution in the United States. The USGS has collected nitrate data over the past 25 years from approximately 124,000 wells nationwide.

Its data revealed areas of groundwater that exceed EPA limits for nitrate in almost every state. The study suggests that a large contributing factor to elevated levels of nitrates is agricultural activities such as the use of fertilizers, irrigational practices, dryland farming, and livestock wastes. Other contributing factors mentioned are septic systems, land disposal of wastes, industrial wastes, and various natural sources. An EPA study of rural water conditions (1986), however, found that only 2.7% of the nations rural wells had nitrate levels exceeding EPA's limits.

In the North Central region, 5.8% and in the Northeast region, 0.3% were reported in violation of standards. Hallberg (1985) determined that since the 1960s, an increase in the use of nitrogen fertilizers has been paralleled by a similar increase of nitrates in groundwater. Fiftyfour million tons of commercial fertilizer were used between 1980 and 1981, 48.7 million tons between 1981 and 1982, and 42.3 million tons between 1982 and 1983 (OTA, 1984).

Table: *Summary of Nitrate-Nitrogen Concentrations in Groundwater, by State*

State	*No. of Wells Sampled*	*% of Wells with NO_3-N 10 mg/l*
Alabama	244	0.0
Alaska	1,305	2.4
Arizona	4,164	13.9
Arkansas	2,436	3.9
California	2,732	10.1
Colorado	5,492	5.7
Connecticut	348	2.3
Delaware	165	9.1
Florida	3,140	2.0
Georgia	1,137	0.5
Hawaii	164	0.0
Idaho	1,806	1.7
Illinois	359	8.4
Indiana	650	1.4
Iowa	4,088	5.0

Kansas	1,140	20.0
Kentucky	3,227	4.2
Louisiana	3,177	0.6
Maine	147	2.0
Maryland	1,215	6.8
Massachusetts	414	1.2
Michigan	1,108	1.1
Minnesota	1,655	9.3
Mississippi	1,701	0.2
Missouri	2,165	2.1
Montana	2,821	3.8
Nebraska	2,326	9.3
Nevada	465	0.9
New Hampshire	69	1.4
New Jersey	1,385	1.4
New Mexico	4,685	2.9
New York	2,491	11.0
North Carolina	908	0.8
North Dakota	7,387	4.6
Ohio	339	2.6
Oklahoma	1,724	11.8
Oregon	685	1.2
Pennsylvania	4,326	5.9
Puerto Rico	79	2.5
Rhode Island	171	36.3
South Carolina	557	0.7
South Dakota	1,996	6.7
Tennessee	109	0.9
Texas	36,196	9.4
Utah	3,301	2.0
Vermont	73	1.4
Virginia	762	0.8
Washington	1,158	4.3
West Virginia	954	0.5
Wisconsin	2,727	3.6
Wyoming	1,477	3.8
Total	123,656	6.4

Source: Madison and Brunett, 1985.

Nitrogen accounts for 6.1% to 20.4% of these weights. Much of the applied nitrate is not utilised by crops and is subsequently lost through leaching. Less nitrate-nitrogen is available for leaching in the loamy finesand soils of subirrigation systems than in furrow and sprinkler systems. Thus, subirrigation systems provide a method for reducing solute concentrations in irrigation return-flow.

Irrigation only contributes further to nitrate leaching. Less than 1% of the total cropland in the Northeast is irrigated; as a result, nitrate contamination is not that much of a problem. In the West, however, due to the highly saline soils, irrigation often exceeds soil requirements.

Dryland farming in the northern Great Plains does not utilise irrigation, and crop plants grown in the area often have lower evapotranspiration rates; therefore water may move down through the soil picking up salts and thus contaminating the groundwater. In some cases this salty water produces saline seeps. In 1983, 12 states experienced groundwater contamination from pesticides applied to agricultural land.

In 1984, a total of 12 pesticides were discovered in the ground waters of 18 states. By 1985, 23 states discovered groundwater contamination from 17 different pesticides. Although agricultural operations alone account for 69%-72% of all pesticide use, contamination by pesticides can also occur through spills, leaks, lawn and garden application, and disposal practices (OTA, 1984).

***Table:** Pesticides Found in Groundwater as a Result of Normal Land Applicationå*

Pesticide	***Use****	***State(s)***	***Typical Positive, ppb***
† Total of 17 different pesticides in a total of 23 different states.			
E* H = herbicide; 1 = insecticide; N = nematicide			
Alachlor	H	MD, IA, NE, PA	0.1-10
Aldicarb (sulfoxide& sulfone)	I, N	AR, AZ, CA, FL, MA, ME, NC, NJ, NY, OR, RI,TX, VA, WA, WI	1-50
Atrazine	H	PA, IA, NE, WI, MD	0.3-3
Bromacil	H	FL	300
Carbofuran	I, N	NY, WI, MD	1-50
Cyanazine	H	IA, PA	0.1-1.0
DBCP	N	AZ, CA, HI, MD, SC	0.02-20
DCPA (and acidproducts)	H	NY	50-700
1,2-Dichloropropane	N	CA, MD, NY, WA	1-50

Dinoseb	H	NY	1-5
Dyfonate	I	IA	0.1
EDS	N	CA, FL, GA, SC, WA, AZ, MA, CT	0.05-20
Metolachlor	H	IA, PA	0.1-0.4
Metribuzin	H	IA	1.0-4.3
Oxamyl	I, N	NY, RI	5-65
Simazine	H	CA, PA, MD	0.2-3.0
1,2,3-Trichloropropane	N (impurity)	CA, HI	0.1-5.0

Source: U.S. EPA, 1986b.

EPA believes that with increased groundwater monitoring more cases of contamination involving additional agricultural chemicals will be discovered (EPA, 1986b). The pesticide dibromochloropropane (DBCP), now banned in the United States, adversely affects human health and causes reproductive dysfunction in laboratory animals when given at high concentrations. There is no direct evidence for adverse effects on human health at the concentrations found in contaminated groundwater (CEQ, 1981a). Although not really an agricultural activity, suburban lawn care can cause groundwater contamination (Miller et al., 1974).

Mining

The formation of acid mine-drainage water has been described above. In areas where coal mining has been carried out over long periods of time, in Appalachia and Pennsylvania, for example, it is estimated that due to its low pH, the groundwater would be unusable for decades after the cessation of mining activities. Metal-mine leachate has increased manganese in wells in Washington, caused arsenic poisoning in cattle in Idaho, and increased the radioactivity of groundwater in Wyoming.

Highway Deicing Salts

Many of the northern states use large quantities of salt in combination with abrasives such as sand on icy and snowy highways. The salt-solution runoff can percolate through adjacent soils or otherwise find its way into the groundwater via storm drains. Due to its low cost, salt is often stored in uncovered piles. Precipitation dissolves the stored salt, which may then infiltrate shallow aquifers. Large amounts of salt are used on northern highways each winter. Over 12 million tons were used in the winter of 1978-79. During 1982 and 1983 at least 9.35 million tons of dry salts and abrasives and an additional 1.78 million gallons of liquid salts were applied to highways (OTA, 1984). During a typical winter approximately 17.6 tons of salts are used per lane per mile (OTA, 1984). The amount of salt used depends, of course, upon the severity of the winter, upon the amount of abrasives used, and upon driver education. Nevertheless, salt usage in the Northeast has resulted in chloride contamination of many drinking-water wells adjacent to highways.

The dangers to public health of high salt concentrations in drinking water are not well known. Water is considered contaminated with chloride when the concentration of chloride exceeds 250 mg/litre; concentrations above that level leave a salty taste in the water. High concentrations of sodium have been linked with hypertension and have been implicated in the exacerbation of certain liver and kidney problems. The American Heart Association has suggested a maximum level of 20 mg/liter in drinking water. High concentrations of salt can also be severely detrimental to plant life.

Atmospheric Contaminants and Acid Rain

Very little is known about the effects of atmospheric contaminants and acid rain on the quality of groundwater. In Michigan chromium in the dust from an industrial plant was leached into an aquifer and contaminated a well. Hubert and Canter (1980b) have reviewed the literature concerning the relation between acid rain and groundwater, and they conclude that acid rain may react with soils and surface waters in such a way as to increase the leaching of metals and nutrients. Thus, the potential for groundwater contamination is present and should be investigated further, especially the potential for lead and cadmium contamination via leaching because of their long environmental residence time.

Figure: *Regional effects of atmospheric contaminants (acid precipitation)*

Infiltration of Surface Water

Aquifers are often in hydraulic connection with bodies of surface water. Under certain conditions, polluted surface water from lakes or streams can percolate into a water-table aquifer, thereby degrading the quality of the groundwater. Also, groundwater development near a body of surface water may draw contaminated water into the aquifer.

Development of Groundwater

Pumping of water wells may bring water of lesser quality into their zone of influence. In inland areas serious problems may be caused by the migration of water from saline aquifers through leaky aquitards into potable supplies. In coastal regions extensive groundwater development has led to saltwater intrusion. Saltwater encroachment is an important source of groundwater contamination in many populated coastal regions, particularly in parts of California, Texas, Louisiana, Delaware, Florida, and New York. Saltwater occurs naturally in the aquifers in coastal areas, and pumping fresh water from wells may cause saltwater intrusion into potable aquifers, rendering them unfit for use.

In 1981 the USGS estimated that 21 billion gallons of water per day, or 26% of all groundwater withdrawn, is in excess of natural recharging capabilities. In 1985 the EPA reported 19 states having problems with saltwater intrusion from deep saline aquifers or from the ocean. Attempts to prevent such contamination include the injection of freshwater "barriers" to reverse the hydraulic gradient in the aquifer so that the flow is toward the sea rather than toward the pumping wells. This method has met with some success in the Los Angeles area. Land subsidence has occurred in some areas of the United States because of excess withdrawal of ground waters. A dramatic example of this occurred over a period of 37 years in San Joaquin Valley, California, where an area of land subsided nearly 29 feet.

Improper Construction and Maintenance of Wells

Faulty, corroded, or ruptured well casings, the linking of two aquifers by a well screen, or open wells can cause contaminants to move from one aquifer to another. Sources of contamination from the three main categories described above have been documented from all parts of the country. Whether they cause localised problems or whether the groundwater-contamination problems fall into larger regional patterns has been a matter of dispute. Our conclusions, based on a study of the documented cases of groundwater contamination available, together with examples of known occurrences of groundwater contamination for selected states. It may be said, however, that groundwater contamination is not usually direct, and in most cases the contaminant must pass through the soil layer. Attenuation of contaminants has been thought to be greater in the unsaturated zone because of the greater possibility of aerobic degradation, adsorption, complexing, and ion exchange of organics and inorganics. Recent investigations on anaerobic metabolism, however, have discovered that many polluting compounds can be significantly degraded under reduced conditions. Dilution, buffering, precipitation, reduction or oxidation, mechanical filtration, volatilisation, sorption, and other processes may operate within the aquifer to purify chemical wastes.

2

Industrial Topography and Groundwater

The corps' counterparts in business have used engineering decisions to follow a rational, economic course, largely divorced from any deference to environmental consequences. Throughout the twentieth century, engineers, in conjunction with construction crews, arguably did more to reshape the physical form of the United States than did any other single group. In particular, they engineered the landscape as they laid transportation lines and built industrial districts.

A common element of engineered structures was their withstanding of environmental forces, particularly water, and their endurance for a considerable period of time. This would have been impossible had engineers lacked a solid grounding in many basic principles of geology and hydrology and had they failed to understand how to employ those principles in predicting the impact on planned projects.

The process of reworking natural topography to suit the needs of an urbanising and industrialising society was refined by 1900. Indeed, as Denis Cosgrove has asserted, "understanding the modern metaphor of a hydrological cycle and the significance of aquifers and ground water were preconditions for large-scale environmental engineering of the modern world" (1990, 5). Draining the fens in England is one classic example, and the infilling of Boston's Back Bay and the waterfront expansion of New York are well-known urban instances in this country. Cutting and filling grades for canals and railroads and building bridges and locks and dams also furthered engineers' knowledge of basic environmental processes. The wisdom accumulated in such projects found its way into a multitude of engineering publications.

The massive body of technical literature available to engineers has been conveniently compressed into a series of textbooks and manuals. This repository represents not the full extent of engineering knowledge but its coalesced essence, and manuals and textbooks provide a basis for tracing the contours of environmental knowledge within the profession - both in terms of adopting concepts in the field and drawing on ideas from related professions. Textbooks contain the basic principles with which most engineers became acquainted during their formative years of training; manuals provide often-updated guidance to practicing engineers and a continuing link to new developments in the field. These two types of essential reference sources were on every practicing engineer's shelf.

Figure: *Identification of groundwater zones depends upon many factors*

Given the extensive and early discussions about controlling surface water and groundwater in the technical literature, recent comments that engineers and other water-management practitioners had little understanding of groundwater hydrology before the 1970s seem at odds with professional accomplishments. In this chapter I examine the attention directed to groundwater by considering what John K. Wright referred to as the "geosophy," or geographical knowledge and philosophy, embodied in the twentieth-century professional-engineering literature on groundwater. To reconstruct the historical contours of engineers' expertise, I look at three elements of practical groundwater knowledge. First, civil engineers constructing roads, railroads, and canals regularly concerned themselves with surface water and groundwater flow. Second, engineers perfected techniques for controlling farmland drainage, along with urban runoff and sewage. Third, they developed methods for adequately protecting other structures from subsurface water. I then illustrate how hydrology and geology entered an engineer's training, how that was put into practice, and the limitations of professional responsibilities to use this knowledge. This reconstruction of environmental expertise is based on a survey of textbooks, engineering manuals, and articles in technical journals. It presents concepts as developed and embraced by the engineering profession by the mid-twentieth century. Where key ideas were well developed before then, older sources illustrate the earlier acceptance of fundamental concepts. Although I cite only a modest number of sources, I drew on a larger sample, and I tracked key sources in reconstructing the engineers' geosophy of water by the 1950s. The chapter is intended not as a defence of engineers' environmental sensitivity or concern but as an assessment of their ability to comprehend environmental process and to design accordingly. This is not to say that this knowledge was always put to use but to demonstrate that knowledge of groundwater and subsurface hydrology was a fundamental part of engineering practice and was relied on to some extent in all circumstances.

Transportation Systems and Water

Figure: *Raw Water Transportation System*

Nineteenth-century canal- and road-building efforts, followed swiftly by railroad construction, forced engineers to come to grips with environmental processes. Each of these endeavours required cutting and filling uneven topography, creating soil or rock foundations for bridges, and providing the most level grade possible for the routes themselves. Fundamental to civil engineering was comprehension of the load-bearing capacity of rocks and soils, the effects of saturation and flowing above ground water on foundations, and the ability of engineered surfaces to withstand the abuse of traffic.

The earliest efforts to engineer the environment amounted to a problem-solving phase worked out in the process of constructing transportation systems. Although the objective of early engineering was not the explicit comprehension of natural systems, as science would approach it, making useful things did require working with the natural medium. To do this successfully demanded substantial environmental knowledge. The National Road, the Erie Canal, and early railroad lines forced engineers to work with the problems of cutting rights-of-way through rugged topography, ensuring that the roadbeds would be adequately drained and not washed away by streams, and also controlling water in canals and behind dams. This phase of environmental understanding was largely a trial-and-error process. As the nineteenth century progressed, understanding became more standardised, and trade associations worked to disseminate information gathered in the field.

Twentieth-century railroad and road building borrowed from engineering techniques developed a century earlier. Engineers hailed straight roads as the ideal form to reduce construction costs, maintenance expenses, and transportation charges. Realising the ideal, however, was often impractical, given existing landforms, and early road-building treatises advised builders to sacrifice distance for a gentle grade.

To obtain a suitable slope through rolling topography, engineers had to select routes where a portion of the road would be cut into hillsides and other segments would be built up with material excavated elsewhere. When road cuts were necessary, standard ratios guided execution as early as the mid-nineteenth century. William Gillespie, author of a nineteenth-century manual, recommended a steep slope (1/4 foot run to i foot rise) in solid rock. For days, he suggested a 45 [degrees] angle (1:1 ratio), although noting that slumping had occurred at a ratio of 6:1. A raised roadbed, even on level ground, was a standard prescription, to assure that the road would remain free of water. Gillespie also stated that "adequate ditches" could transform a bad road into a good one (1847, 53).

This guidance remained a fundamental doctrine of transportation engineering into the twentieth century. In his road-building treatise Ira Baker, a civil-engineering professor at the University of Illinois, drew on the numerous publications that issued from the many state highway departments (1914). At a formative stage in highway-building technique, Baker expressed an obvious concern with water-related problems and demonstrated the level of communication among practitioners. He insisted that "drainage is the most important matter to be considered in the construction of roads, since no road, whether earth or stone, can long remain without it". He pointed out three essential elements of a satisfactory drainage system: underdrainage, to remove standing water within 3 feet of the earth's surface; surface drainage, or crowning the road surface to keep water from penetrating the road and softening the surface; and side ditches, to collect and transport water away from the road as quickly as possible. Where roads had to be cut into the earth, he recommended shallow ditches underlain by drainage tiles. These devices functioned in lieu of larger ditches that would have been possible on flat land with expansive rights-of-way.

In some situations, limited amounts of water could prove desirable. Baker pointed out that moisture would bind loose grains in a sandy road and thereby improve it. But moisture damaged clay roads. To achieve the best properties of both clay and sand he advocated mixing coarse sand or gravel with clay to form a hard, nearly impervious surface that water would not readily soften. In order to select a satisfactory clay-sand mix, Baker described tests to determine the relative percentages of clay and sand. Soil evaluations were fundamental to the creation of a roadbed with the desired properties (1914).

Railroad-engineering textbooks discussed concepts relating to slope stability and water. Walter Loring Webb's classic treatise on railroad construction (1932) is an example of the transfer of information directly from the leading railroad-engineering organisation to the engineering classroom. It indicated that the slope for rock cuts varied depending on the type of rock. Webb pointed out that hard rock, which would not disintegrate when exposed to weather, could stand with a nearly vertical face but that a gentle slope (4:1 run-to-rise ratio) was necessary for soft material such as clayey soil. For rock fill, Webb noted that a steeper angle (1:1 ratio) was satisfactory, but for earth fills a gentler slope (1 1/2:1) was advisable.

This basic engineering practice reflected general principles of hydrology: namely, that hard rock was more resistant to erosion than was unconsolidated material and that it would hold a steeper slope. Additional hydrologic knowledge appeared in the section on procedures for finishing a grade. Webb repeated the American Railway Engineering Association (AREA) specifications that called for a subgrade with a crowned surface, "thus leaving no depression that will hold water". He reprinted an illustration from the 1929 AREA manual that showed drainage ditches paralleling roadbeds. The manual asserted that "water should be kept off the roadbed if possible". The drainage system it illustrated included interceptor ditches upslope from the roadbed, ditches with drainpipes adjacent to the roadbed, and another drainage ditch beneath the toe of the roadbed fill. Recommended engineering methods embodied an awareness that soils could become saturated and fail but that drainage systems could capture and transport both surface water and groundwater away from the roadbed.

Where waterways intersected a roadbed, railroad engineers had to design passageways beneath the track, and engineering textbooks again provided guidance. Webb expressed it this way: "Although a variable percentage of rain falling on any section of country soaks into the ground and does not immediately reappear, a very large percentage flows over the surface, always seeking and following the lowest channels. In order to prevent injury to railroad embankments by the impounding of such rainfall, it is necessary to construct waterways through the embankment". A properly designed culvert would not allow water to back up and saturate the embankment; to execute a satisfactory design required knowledge of local precipitation patterns, of the size and slope of the watershed, and of soils and vegetation. Each of these factors affected the movement of water and determined how quickly and in what quantity it would arrive at the passageway beneath the track. This knowledge had been worked out by hydrologists and readily accepted by railroad engineers by the nineteenth century and underwent refinement during the twentieth century.

Transportation engineers had to understand the dynamics of groundwater to ensure adequate water supplies in arid regions. Water was an essential ingredient for steam locomotives, and where tracks crossed arid regions, groundwater often was the sole available source. Knowledge developed and disseminated by railroad engineers, for example, started with how to locate and select an adequate and safe water supply - requiring geological and chemical evaluations - and proceeded on to methods for drilling and finishing wells and the techniques for protecting a water supply. Technical publications specifically advised railroad crews to be aware of the potential for incidental groundwater contamination resulting from careless disposal of wastes or poor well-finishing techniques.

The 1929 AREA manual even warned that inadequately controlled fuel oil could damage neighbouring properties. Additionally, railroads were concerned with water quality and had to avoid hard water that would encrust a locomotive's boiler. This required the ability to analyse water quality, and the railroads relied on existing techniques to test for hardness, various metals, and sanitary condition. Overall, transportation-systems engineers focused considerable attention on the interaction of their structures with the local environment, particularly hydrologic systems. Moisture was a persistent cause of concern, but one that could be managed with proper design. Engineers sought to collect and remove water, whether surface or subterranean, and used the basic forces of hydrology to accomplish their goal.

Controlling Water on Farms and in Cities

The control of water in fields, in streams, and under structures has been another critical engineering task. Efforts undertaken to drain the massive wetlands in the Midwest reflect hydrologic knowledge within both professional and folk communities. At the most fundamental level, where there was excess water, farmers could install drainage systems to dry the land. Ditches that transported water from saturated fields proved satisfactory in the early nineteenth century; use of buried tiles was an improvement later in that century. Not only did tiles drain the fields, they did not consume cropland. In addition to the rural experience with subsurface drainage, sanitary engineers began to employ citywide drainage systems to remove both sewage and storm runoff during the late nineteenth and early twentieth centuries.

Figure: *Water for Cities, Towns, Farms & Animals*

First compiled in 1908, Thaddeus Merriman's American Civil Engineers' Handbook covers the breadth of civil engineering and contains extensive treatment of field and city drainage. Farm drainage afforded farmers several benefits: It removed surplus water and aerated the soil; it warmed the soil and decreased the risk of frost; and it increased the friability and moisture-absorbing properties of the soft. Although the last benefit may have been contrary to the original objective, these three factors represent the prevailing understanding of water removal. To successfully achieve the benefits, engineers first had to analyse soil and groundwater properties. The manual advised that soil and topography had to be mapped and that "the ground water also should be studied, especially if there is evidence that it comes from some source other than precipitation upon the area to be drained [regional groundwater flow]". There was concern with both surface and subsurface waters. Furthermore, the manual describes subsurface drainage as the removal of water that cannot be retained in the soil. Drains for transporting subsurface water were to be buried in the fields "with openings of a size that will permit the ground water to enter freely and will exclude the surrounding earth". Designing the capacity for the drainage system depended on soil conditions and precipitation, and because field drainage was such a costly endeavour, the design had to be correct the first time. Known conditions enabled the prediction of future outcomes.

Similarly, engineers who planned urban drainage systems had to design sewers that would accommodate storm runoff, sewage discharges, and infiltrating groundwater. Typically, eastern and midwestern cities employed combined municipal sewers to handle both storm and sanitary sewage. Designing such a system entailed calculating storm runoff. Elements of the formula used by engineers included the drainage areas served by the sewer, the intensity of precipitation planned for, and the coefficient of runoff. To determine intensity, a twenty year rainfall record was considered necessary to calculate the frequency of extreme precipitation events. In determining the coefficient of runoff, engineers had to make judgments not only on what to use as the extreme event to design for but also on how to estimate the pervious and impervious surfaces within the drainage basin. To design for domestic sewage, the manual offered averages in the range of 200 to 250 gallons per day per capita but recommended local data collection for industrial waste volumes. Projecting population growth played a large role in the design. Finally, engineers had to consider the amount of groundwater that infiltrated into sewer systems. Even in well-designed and -maintained systems, groundwater inevitably found its way into sewers-with, perhaps, exceptions in arid regions of the West - and the author reported that 50,000 gallons per mile per day were not uncommon. The manual recommended that engineers consider seasonal fluctuations in groundwater volume and the water-table level, or not to assume the presence of a static water table. The ultimate design had to accommodate all three components and involved the elaborate prediction of both surface water and groundwater movements.

Figure: *Water is the newest cash crop for some farmers in California, with cities eager to buy it. Image from Moffett Field*

Rapid urbanisation in the post-World War II period presented problems for public-works engineers. Conversion of once-pervious surfaces into streets, houses, and parking lots rapidly increased storm runoff and forced engineers to re-evaluate formulas for handling increased volumes of sewer discharges. In the late 1950s and early 1960s the U.S. Geological Survey (USGS) conducted several studies of urban hydrology and proposed remedies for controlling surges of storm runoff. The belated recognition of the problem suggests that despite prescriptions that called for a detailed and prolonged assessment of local hydrologic regimes, engineers sometimes ignored prevailing guidelines or used outdated information.

From the rather informal engineering of small earthen dams to provide hydropower for gristmills to the construction of such massive structures as Hoover Dam, builders had to contend with the pressures of water against retention structures and provide features to prevent seepage that would undermine the structure. Designers of dams - and similar structures, such as levees and dikes - employed formulas to identify the path of potential seepage and modified their designs to prevent it. And cases of dam failure, such as the one that caused the Johnstown flood, were well known.

At the heart of such design work was a basic engineering calculation. The Darcy formula described the relationship among pressure (hydraulic gradient), soil conditions (permeability), and discharge velocity - velocity equals the constant of permeability times the difference of head (hydraulic pressure) divided by length of column. It served as a forecasting tool to predict the yield of a fluid under a known pressure based on the Darcy formula and other calculations, the basic physics of groundwater flow indicated that moisture would flow underneath an earthen dam or embankment in elliptical paths.

Beyond subterranean flow, moisture would saturate a dam in a line descending from the point of contact downward toward the opposite side of the dam. To decrease the velocity of water percolating through an embankment, engineers used the relationship presented by the Darcy formula, namely that velocity could be decreased by lengthening the distance. This could be achieved by placing an impervious blanket at the inner toe of the dam or by inserting sheet piling that extended below the base of the dam. Either procedure elongated the distance that water had to travel and thereby reduced its velocity and prevented sufficient flow to erode the base of the embankment.

When carrying out construction projects below either the surface water level or the water table, engineers had to drain the construction site. For dams and bridges or pier pilings, engineers used coffer dams to create a setting where work could proceed beneath the water level. Several related techniques permitted this. In still water, workers dumped earthen material in the form of a dike around the construction site and pumped the water out. Other situations, particularly with currents, necessitated more elaborate structures. Crews could drive sets of parallel wooden piles into the stream or harbour bottom around the site and then place sheet-piling walls between the timber anchors, sometimes filled with a mud "puddle" for strength. A third method used on bedrock featured a sheet-metal coffer wall backed by a timber and rock-filled crib for stability. Each of these methods took into account the base material and created a barrier between the work site and the standing water.

Engineers accepted leakage through and beneath coffer dams as an expected outcome of these temporary features. Walter Douglas, in a chapter in Merriman's opus, stated that "the object of a coffer dam is not to prevent all infiltration, but only to reduce it to the extent that a moderate amount of pumping will keep the water out of the way" (1930, 752). To deal with the water that entered the construction site, manuals recommended sump pumps to collect and lift the water back out.

When foundations merely encountered the water table, ditches served to drain the area and lower the upper level of groundwater. If the foundation extended deep into the earth and surface drainage would not remove sufficient quantities of water, well points might provide drainage. Well points are wells drilled around a construction site. They take advantage of the concept known as the cone of depression; by removing groundwater faster than recharge takes place, they lower the groundwater in a cone-shaped area around the well. With the installation of several wells, engineers could lower the water table under a construction site by creating coalescing cones of depression. By the late 1940s specialised companies were carrying out this procedure, which required a combined knowledge of the depth of the aquifer to be drained, its recharge rate, and the permeability of the earth material.

Standard volumes such as Clarence Dunham's engineering textbook provided detailed guidance for the construction of waterproof foundations (1950). Preparing for waterproofing, engineers typically assessed seasonal water-table levels. Beyond that, various concrete mixtures could slow leakage by reducing the porosity of the concrete. Nevertheless, no concrete mixture could stop all movement of water through joints or cracks. Impermeable coatings might reduce leaks, but Dunham asserted that coatings were a last resort and of dubious efficacy. He claimed that "it is best to prevent leakage by proper planning in the original design". A well-planned foundation would have drainage lines - similar to those used in agricultural fields - to remove water. Dunham recommended multiple layers of water-resistant membranes, combined with a drainage system, and water-resistant jointing.

Storage of fluids in canals, reservoirs, or lagoons was a further everyday aspect of managing water resources. In laying down rules for reservoir design, manuals insisted on an initial survey of the drainage basin, to include the geology of the reservoir site, the availability of construction materials, and the relative cost and value of water storage. Once a site was selected, a detailed geological investigation to anticipate and forestall seepage loss was essential. As early as 1918 an article on the construction of reservoirs and canals warned against selecting sites overlying gypsum or limestone deposits, coarse-grained sandstone, or the other coarse materials that would permit escape of impounded fluids. Later Merriman would warn that "many instances can be cited where, after construction, reservoirs have proved worthless because of their inability to hold water" (1930, 1673). His manual pointed out that seepage losses were reduced where rock or clay foundations existed and grew where the underlying material was loam or volcanic soil. Recognising the inconsistency of subsurface material, Merriman cautioned: "If any portion of the banks or bottom of the reservoir is formed of sand or gravel even though it may be overlain with a blanket of fine soil, and such sand or gravel might form an underground channel leading around or under the impounding dam or to some other drainage basin, there is grave danger of excessive seepage loss".

Works on soil mechanics pointed out that percolation was one of two principal causes of water loss in canals. They reported that moisture escaping a canal could join groundwater and ultimately move with it, generally destined toward a local surface stream. Determinations about the overall benefits of a projected canal project hinged on the ability of the canal to contain water and not to lose more than it transported. There were basic field methods for determining permeability that allowed engineers to estimate the projected water loss. Such calculations enabled them to recommend against expensive canal projects if the costs did not justify the amount of water delivered.

Lagoons, like reservoirs and canals, were ineffective if they did not retain fluids as desired. Yet lagoons had numerous industrial uses, from storing process water or other fluid raw material, to temporary storage of industrial wastes, to treating effluents. In the 1940s E. F. Eldridge advised that lagoons containing toxic wastes could create a menace to wells in the vicinity by way of seepage (1942). To prevent seepage that would lead to groundwater contamination, he recommended lining impoundments with concrete. Subsequently, a respected sanitary-engineering treatise advised that "lagoons of any type must be located on a site where seepage will not contaminate wells or underground sources of water and must be constructed with watertight embankments".

Engineers who worked with lagoons understood the imperfect nature of liners that were supposed to prevent leakage. Natural clays were recognised as potential liners, and in the 1940s it was not uncommon for engineers to recommend that clay liners be constructed as part of the lagoon. But such features had their limitations. Dimitri Krynine, in a discussion of soil permeability, pointed out that the bonds of undisturbed clays were stronger than were those of disturbed clay, such as transported and compacted soils (1941). The chemical reaction of the material in the lagoon with the liner was another potential problem. One waste-treatment expert cited a case in which chemicals in a lagoon had reacted with the liner to cause its failure (Powell 1954). Although lagoons sometimes failed to function as intended, the prevailing professional view was that design modifications could effect a desired performance, whether containment or infiltration.

Drawing on Geology and Hydrology

Intensive groundwater pumping by English cities had caused serious problems for nineteenth-century farmers whose springs and wells dried up as a result. Debates over water rights brought geologists and engineers together as consultants for the contending parties, and they investigated the relationship among groundwater, pumping, and surface land uses. Although the remedies for problems "were often inept and piecemeal", the ruling order among engineers nonetheless built on the hydrologic foundation of previous generations, adding to the formal canon and further expanding comprehension of subterranean waters for subsequent researchers. Communication between engineers and geologists continued as a critical professional link into the twentieth century. Bedrock mining had been limited to above-groundwater depths before the mid-nineteenth century, except in rare situations where water could be drained off by gravity or where barriers to seepage could be constructed. With advances in pumping technology, mining below the water table became fairly common by 1850. By the turn of the twentieth century, mining engineers considered the Cornish Pump the most reliable among an array of available devices. It could lift hundreds of gallons of water out of shaft mines with little risk of breakdown.

Figure : *Concerns arising from the Geology and Hydrology of the ground*

John Schultz and Arthur Cleaves summed up the long, historical importance of groundwater to engineers. In 1955 their engineering geology textbook they wrote: "If one were to select the branch of geology most important to civil engineering, it would probably be subsurface water. Few, if any, engineering projects are entirely unconcerned with this subject" (1955, 93). They further pointed out that groundwater was a fundamental concern in constructing dams, tunnels, reservoirs, irrigation systems, and building foundations. Engineering American industrial topography required sophistication with groundwater processes. The preparation of a textbook for engineers by a pair of geologists underscores the co-dependency of the two disciplines, a professional exchange that began long before the mid-1950s. Hydraulics was a standard part of a civil engineer's training, and drainage was a central component of one's education (Seely 1993). Although the early prominence of groundwater studies in engineering and geology may seem obvious, there are nonetheless recent claims that no formal training in groundwater processes was available until after 1965. That engineers drew on related scientific literature is aptly illustrated by the depiction of groundwater in the essential civil-engineering textbooks. Education in engineering was essentially and persistently cross-disciplinary. In a 1939 book, Geology and Engineering, Robert Legget argued that scientific groundwater studies were born as "modern engineering" began in the nineteenth century. According to Legget, the English civil engineer William Smith investigated the relationship between geology and groundwater and cured engineering problems by applying his findings. In particular, he developed a water-supply system for Scarborough and drained areas near Bath as evidence of his knowledge of groundwater principles.

Legget illustrated his prewar textbook with a figure of groundwater zones reproduced directly from a uses publication. It shows the upper zone of aeration, where water from precipitation seeps into the soil and, under the pull of gravity, moves downward. He differentiated between vadose, or soil water in the upper zone that is not saturated, and the zone of saturation or groundwater. The basic hydrologic concepts that fluids from the surface move downward to a zone of saturation and that groundwater lay within this zone of saturation were accepted as tenets of basic engineering knowledge in the 1930s. Interdisciplinary communication is clear in the use of nearly identical groundwater illustrations that would appear in engineering books for the next two decades and that provided engineers with their introduction to the subject. Furthermore, in their general treatment of engineering geology, Schultz and Cleaves (1955) refer their readers back to O. E. Meinzer's pioneering work at the uses and to C. F. Tolman's standard groundwater treatise.

The borrowing and exchange between engineers and geologists appears later in the geotechnical textbook by Dimitri Krynine and William Judd. The illustration of subsurface water zones in their section on "Subsurface Water" is straight from a basic geology tome. Additionally, several sketches of groundwater movement, most notably groundwater mounds, come directly from Tolman's Ground Water (1937). Krynine and Judd note that engineers generally dwelled on the unfavourable effects of groundwater on foundations and structures (1957). In a discussion of groundwater investigations, they point out the ability of engineers to determine the location (depth) of the water table and the velocity and volume of groundwater. Their references rely on material produced by the uses and the American Society of Civil Engineers.

Engineers accepted several basic principles of groundwater flow and put them to use in various forms of water storage. Legget discussed the process of underground dam construction and water impoundment. These subsurface structures impeded groundwater, allowing it to accumulate behind the barrier. He cited numerous examples in which such storage systems were used as a reservoir for public water supplies (1939). Groundwater-recharge projects in California and Illinois demonstrated before 1940 that lagoons with permeable bottoms could be used to recharge aquifers beneath them (Tolman 1937). Such activities are just a start among the practical applications of groundwater theory.

Engineering Industrial Sites

Natural topography was seldom satisfactory for the large, land-extensive industrial facilities that were built around 1900. Construction of such facilities reflected both the comprehension of hydrologic knowledge, which made them practical, and the confidence that information derived from analysis of conditions at diverse locations would yield the data needed for engineering formulas. The conversion of the Lake Calumet wetlands on the southeast side of Chicago into a vast industrial complex required extensive filling of waterfront areas and a thorough dredging of the slow-flowing waterways near the southern shores of Lake Michigan.

In 1878 the North Chicago Rolling Mill began to extend the shoreline near the mouth of the Calumet River, using slag and refuse as fill. The mix of permeable material permitted ready drainage of the site. On the west bank of the shallow Lake Calumet, George Pullman created his town and industrial site by raising the level of the land 5 feet above the surrounding marsh with material dredged from the lake bed. Fine-grained lake-bed sediments did not drain as well as slag, so Pullman installed an extensive sewer system to carry water away from his model town. By the first decade of the twentieth century, another 400 acre extension of land south of the Calumet River's mouth housed the Iroquois Iron and the American Smelting and Refining Company facilities. Much of this extension was built on slag (Colten 1985). The coarse residue from iron drained quickly and acted as a sink for liquid waste. A massive lagoon received effluent and allowed sediments to settle out and liquids to percolate into the fill. When industrial engineers went to work constructing the Jones and Laughlin Aliquippa works near Pittsburgh, the contractors laid out a factory site along the Ohio River. Much of the property was below the high-water mark. This was not taken as an obstacle, however. According to a contemporary account, the low level "offered an excellent dumping ground for blast-furnace slag, flue dust and cinders, and it has been steadily filled until a large part of it is now above high water mark". This brief comment quickly captures several elements of engineering knowledge of the environment. First, builders evidently were aware of flood records and found it advantageous to fill the site with their own wastes, reducing the risk of floods. The slag and cinders provided excellent drainage and could serve as adequate foundations for future expansion. The site engineers took flooding as well as waste disposal into consideration as early as 1913. Ultimately, the Jones and Laughlin site came to resemble a broad floodplain terrace, but one consisting of coarse slag and other fill rather than fine-grained sediments. This provided superior drainage and stability for the steel-mill structures.

When U.S. Steel built a facility in Gary, Indiana, it selected a site amid dunes and swamps along the shore of Lake Michigan. Construction began in 1906 with crews leveling the dunes and filling the wetlands. The sandy soils provided both excellent drainage and an adequate foundation for the steelworks. Industrial companies expanded their operations into Lake Michigan, using slag as fill. Excavation of sand for roadway construction created numerous pits in Gary's western neighbourhood; use of these sand pits by waste-disposal firms created a lasting legacy of hazardous wastes in highly permeable soils.

The expansive wetland known as the American Bottoms, east of St. Louis, experienced extensive modification for industrial operations. Most of the major metalworking operations used slag to raise their sites. Granite City Steel smoothed out low areas of its property with solid waste. The St. Louis Smelting Company partially filled a small pond with lead slag and spread the remainder on its hilly property. Hundreds of railcars of slag went into the construction of levees surrounding the industrial district on the American Bottoms. The levees protected the industrial district from Mississippi River floodwaters but had to be supplemented with a massive drainage system to divert excess runoff from the surrounding uplands. In addition to filling and diversion works, massive pumping of groundwater lowered the water table and effectively drained numerous lakes in the area. This made possible industrial and residential expansion into areas that had been waterlogged. In effect, the combined techniques of filling and drainage served to rework the utility of the American Bottoms between 1890 and 1950.

Such undertakings were not haphazard; they were designed and executed under the direction of engineers. With fundamental training in soils, hydraulics, and foundations, engineers, especially in the Northeast and the Midwest, executed projects with similar goals in diverse settings. For a massive steel mill to stand erect on a heap of slag or on a reworked sand dune, there had to be expert sampling of local geology and the design of structures to serve clients. Crumbling during the first wet season or freeze-thaw cycle was definitely undesirable. To safeguard their work, engineers had to understand in detail any environmental processes that would affect their project. Engineering may have been a trial-and-error process, but once errors in basic procedures or designs were detected, the profession worked as one to craft standards of practice that would elevate all engineering to a safer level. From that standard, experimentation and creation would improve methods and advance results.

The Engineer's Geography

Engineers operate at different scales. An engineer's primary geographical concern is with the interaction between a structure and the geology and hydrology of its site. If a project occupies just one lot, the geography might be constrained to that property. But for a chain of dams on a river system, the geographical considerations expand to the boundaries of a watershed. Regardless of the scale of a project, regional climatic information is important. As a consequence, local and regional information are merged.

Beneath the footprint of a planned structure, shallow core samples or trenches revealed pertinent information for many projects. For structures with excessive loads, a deeper reach was essential to assess geologic and groundwater conditions (ASCE 1934). The engineer's geography may have been limited in horizontal perspective, but it did require details from well beneath the surface. Commonly this entailed the review of regional groundwater and geologic reports and maps. Although a project required localised information, it was available in a regional format, and practitioners had to know the conditions at their work sites. Roads and railroads traversed tremendous distances, crossing a variety of soil and rock types, with relatively narrow rights-of-way. The engineer's geography was a constricted path, relying on information compiled at the county or state level. Geologic and topographic maps guided route selection and familiarised the designer with multiple options, as with railroad routing (Dodge 1965; Peyton 1996). Although it was not necessary to know the source or ultimate destination of streams that crossed the route, their presence, relative size, and discharge mattered. A road-builder's horizon may have been relatively narrow, but it extended well beyond an immediate right-of-way. Likewise, building a levee system over a great distance compelled evaluation of the ability of the local substrate to support massive earthworks and of how water might be transmitted beneath the structure. It was known that permeable lenses of sand offered a path for water to pass under the levee, and techniques were available to assess the potential for sand boils, to interrupt the undermining flow. Engineering municipal public-works systems called for a more comprehensive understanding of environmental processes. Sewerage systems that carried both storm runoff and sewage had to be fitted to cities. Engineers would calculate runoff for large surface areas and add expected discharges from homes and commercial operations. Watersheds and stream systems were the objects of investigations at the Corps of Engineers' Waterways Experiment Station, where mock-ups of river segments allowed experimentation with possible control structures. Engineering had to consider regional geology and hydrology and to draw on national studies. If engineers showed little concern for the effects of their constructed creatures beyond rights-of-way or municipal limits, they were most surely aware of what impinged on their projects from beyond their particular boundaries. Incorporating regional environmental knowledge, they were able to design and construct with an expectation of long-term structural integrity.

Failure to apply known techniques to prevent adverse outcomes was not unknown. Water percolating beneath dams and levees undermined them in numerous floods. Industrial wastes seeping from lagoons into the groundwater has occurred all too frequently. Engineering expertise made such outcomes predictable in the 1950s.

The Engineer's Way

At a fundamental level, the engineering of industrial topography was a predictive art. Beginning with a trial-and-error phase in which practitioners collected field data, it analysed relationships between natural processes and engineered works. Using this basic information, engineers employed various rote formulas, scribed in manuals and textbooks that were part of an engineering canon, updated with experience, to calculate known physical relationships in particular circumstances.

For each new project, local information had to be collected and specifics inserted into the formulas. Based on the site-specific information and general processes, engineers would predict storm surges, runoff rates, or seepage paths. With this knowledge as a first principle, they could design roadways, canals, building foundations, dams, and lagoons to withstand the anticipated environmental forces. Typically, multiple design options were available to accommodate the varying conditions. Such was the art of engineering - selecting the right solution for a given problem - an art that required consideration of long-term interactions between a construction project and its environmental setting.

To gain added expertise that would further help them deal with environmental problems, engineers took courses in soils, hydraulics, hydrology, and geology. Curriculum offerings in these subjects were a standard part of the civil engineer's ongoing education. Textbooks introduced engineers to the findings of the USGS and other experts who conducted research in surface water and groundwater hydrology; as in law and business education today, case studies were a basis for study. Manuals continued exposure to the basic concepts and to new developments in related fields. Furthermore, engineering that required detailed examination of local conditions at building sites called for direct communication with geologists and hydrologists. At every level of education and practice, engineers had to cipher in the interaction of their projects and local conditions.

Whereas the scale of drainage and much engineering in cities was metropolitan, work on agricultural fields, canals, railroads, and industrial sites was often extremely site-specific - but even there, engineers extended their reach to larger regional contexts. That they failed to use the full arsenal of knowledge in some cases indicates the constraints of budgets, standardised design criteria, and other factors. The infusion of environmental concerns into engineering practice in recent years has broadened perspectives, but without introducing any fundamental information about environmental processes or ecology. The professional literature abundantly demonstrates a geosophy that included groundwater and its processes but that missed much of the natural world that was not an active threat to concrete or construction.

An Introductory Hydrogeology Course

In a senior-level hydrogeology class, teams of students conduct a semester-long design project of a hydraulic containment system for a hypothetical contaminated site. This project provides a practical application of topics in the course, and it serves as a better vehicle tnan traditional homework, problem sets, etc., for integrating the various topics of the course. Traditional homework assignments typically contain stand-alone problems or groups of problems that are isolated from other assignments of the course. In this design project, students characterise the hydrogeology of the contaminated site by building on topics covered throughout the course; thus, many topics from the course are integrated into one long-term learning activity.

The class is comprised of senior-level civil engineering students and graduate students in civil engineering, geography, and geology. The typical student in the class has a solid background in mathematics, some experience in higher-level computer programming (e.g., Matlab, Visual Basic) and basic numerical methods, but little or no exposure to geology. Most students have completed an Engineering Hydrology course that includes one to two weeks of groundwater hydrology.

The project involves a hypothetical spill of trichloroethylene (TCE) in a hypothetical aquifer loosely based on a region of Socorro County, New Mexico. I prepared a MODFLOW-2000 model of the hypothetical aquifer, and performed various simulations to obtain hypothetical data. Students receive all of the data at the beginning of the semester. Throughout the semester, they perform various activities such as drawing a water table map, calculating the hydraulic gradient, calculating hydraulic conductivity based on permeameter tests, slug tests, pumping test, and analysing well logs to determine aquifer thickness, as these topics are covered in the course. The final step of the project is to use their aquifer parameters in the design of a hydraulic containment system for the hypothetical spill site. There are several benefits of the project. The students have the opportunity to apply the course material to the simulation of a real-world problem. In addition, the students see the interconnectedness of the course topics. Many of the topics can appear disjoint; however, all can be used in a groundwater investigation. Another benefit of the project is that students learn how to use sparse data sets, which are common in hydrogeology applications, hi this project, hydraulic conductivity and head are known only at a few locations, and the students must determine how to use these data to obtain estimates of aquifer parameters. In addition, the hydraulic conductivity measurements are made on three different measurement scales (cubic centimeters for permeameter tests, cubic meters for slug tests, and on me order of thousands of cubic meters for the pumping test).

Through the project, students learn to discriminate among data measured at multiple scales. A final benefit of the project is enhanced learning from group discussions with team members. The purpose or this chapter is to introduce the project, and to demonstrate its effectiveness in enhancing student learning in an introductory hydrogeology course. The project handouts, model input and output files, and the codes used to generate the data sets are available from the author. The project scenario is presented in the next section.

Project Scenario

A truck carrying trichloroethylene (TCE) has overturned on an exit ramp off of Interstate 1-25 near Escondida in Socorro County, New Mexico, spilling much of its contents onto the land surface. After clean-up, it was estimated that approximately 2 kg of TCE was unrecovered and assumed to nave infiltrated into the subsurface. The spill site is identified as approximately a 61 m X 61 m area shown. The community of Florida, which is situated approximately 0.8 km south of from the spill site, operates two pumping wells that provide water for approximately 60 households. The well locations are labelled FLl and FL2. FLl is pumped at a rate of 42.5 m^3/d, and FL2 is pumped at a rate of 33.9 m^3/d. Both wells were installed in July 2003, and have been pumping continuously since August 2003.

The aquifer is phreatic, with water flowing approximately from north to south, approximately paralleling the Rio Grande. The aquifer is relatively homogeneous, except along an arroyo where hydraulic conductivity is lower than the surrounding aquifer.

Project Requirements

To prevent contamination from reaching the pumping wells FL1 and FL2, the students must design a two-well hydraulic containment system to contain the TCE within the spill site. The containment system should have a pumping well downgradient of the spill site and an injection well upgradient of the spill site, where the pumped water will be reinjected. The design parameters include well locations, pumping and injection rates, and the screened interval of pumping and injection wells.

If the pumping and injection wells are aligned in the flow direction, a contained zone will develop that is hydraulically isolated from the rest of the aquifer. The size of this contained zone depends on pumping and injection rates, distance between the pumping arid injection wells, aquifer type (confined or unconnned), aquifer thickness, hydraulic conductivity, and ambient hydraulic gradient. Thus, the first part of the project involves data analysis to determine the aquifer properties and the hydraulic gradient. The second part of the project involves using tnese property values to aid in the selection of well locations and pumping rates. These parameters should be chosen so that the contained zone includes the entire spill site.

Aquifer Parameters - To estimate the hydraulic gradient, students are provided with water levels measurements at each of the wells and piezometers. These water levels are influenced by the ambient flow conditions, and by pumping at FLl and FL2. Students draw a water table contour map for the area, and from that, they estimate the magnitude and direction of the hydraulic gradient near the spill site.

To determine the aquifer type and aquifer thickness, students are provided with well logs from the two pumping wefls (FLl, FL2) and the six test wells (TW1-TW6). The well logs were recorded during drilling; therefore the water levels in the well correspond to the ambient conditions, prior to any pumping at FLl and FL2, and are higher than the water levels. All well logs show a relatively uniform material type from the land surface to the bedrock, indicating that the aquifer is phreatic. The well logs show the land surface elevation and the depth to bedrock. Using this information and the water levels, students can determine the saturated thickness of the aquifer at these eight locations.

Using results from permeameter tests, slug tests, and the pumping test, students determine hydraulic conductivity at several different locations near the spill site. Except for the data from TW2, the measured values from hydraulic conductivity are all within an order of magnitude; however, a single value must be chosen to be representative of the spill site. Students choose a representative value while considering the proximity of the measurement location to the spill site and the sampled volume for each type of test. In addition, they also select a representative aquifer thickness.

Although the aquifer is mildly heterogeneous, and spatial data are available for hydraulic conductivity and aquifer thickness, the students select a single, representative value of transmissivity to use in their design. There are two reasons for this. First, the sparsity of the data, combined with the correlation structure of the random distributions of hydraulic conductivity and aquifer thickness, limit our ability to characterise the spatial correlation structure of these properties on a scale of the size of the contained zone. The spill site covers an area of 61 m X 61 m, and the contained zone typically cover an elliptical area with a long axis that is approximately double the length of the spill site (on the order of 120 m). The random fields that represent the hydraulic conductivity and aquifer thickness have a correlation length of 304.8 m. This correlation length is on the order of the separation distance between the locations at which hydraulic conductivity measurements are available. Thus, because of the coarse spacing of the data, a variogram analysis of the data would not identify the correlation structure of the underlying spatial distribution of hydraulic conductivity. In addition, the typical length of the contained zone is smaller than the spacing between measurements; thus no information regarding the spatial variability within the contained zone can be obtained. Secondly, although we discuss heterogeneity in this class, we do not cover methods for characterising spatial correlation.

Design of a Hydraulic Containment System - Once the aquifer parameters are identified, students design their hydraulic containment system. The wells are aligned with the direction of the hydraulic gradient, with the pumping and injection wells being equidistant from the centre of the spill site. The wells are designed to be fully-penetrating to ensure complete containment in the vertical direction. The remaining design parameters are the pumping and injection rates, Q, and the distance, L, from the pumping well to the injection well.

Students set up a MODFLOW-2000 simulation to test various combinations of flow rate and distance L to determine which combinations result in hydraulic containment of the spill site. The aquifer is modelled as a one-layer unconfined aquifer with a thickness equal to the representative aquifer thickness. The domain is rectangular, aligned in the direction of flow, as determined from the hydraulic gradient. Specified head boundary conditions are used on two opposite boundaries to create an ambient flow field with the modelled hydraulic gradient set equal to the hydraulic gradient determined from the water table contour map. No-flow boundary conditions are used on the remaining two boundaries. The aquifer is treated as homogeneous with a hydraulic conductivity equal to the representative value of hydraulic conductivity. Steady state simulations are run for various combinations of Q and L, and head contours and flow paths are drawn to determine whether or not the spill site is contained.

The MODFLOW-2000 model is set up using the graphical user interface software, PMWIN. To prepare students for the modelling component of the project, one class period during the fifth week of the semester is used for a PMWIN and MODFLOW-2000 tutorial in which the students complete the tutorial exercise in Chiang (2005). At this point in the semester, students have been introduced to the governing equations of groundwater flow and have derived some analytical solutions for simple cases. Prior to the computer lab exercise, I briefly introduce the finite difference method, emphasising the spatial discretization and the need for spatial data to be defined in each finite difference grid block. With this background, they understand the need for the various inputs into MODFLOW. This tutorial is followed by a homework assignment in which the students use PMWIN to set up a simple MODFLOW model on their own. In the tenth week of the semester, at the end of the unit on well hydraulics, which includes superposition, students do another in-class computer lab exercise using MODFLOW to simulate a hydraulic containment system for predefined conditions, which is then graded as a homework problem. After completing both computer lab exercises and related homework assignments, students have sufficient experience and confidence in running MODFLOW simulations.

The final design must contain the spill site and must minimise power required for pumping. The total cost of the design includes construction costs, drilling costs, and operating costs. It is assumed that the only cost that depends on the design Q and L is the cost of pumping; therefore the design that minimises the pumping power requirements willalso minimise the total cost.

Model and Simulations

The aquifer is phreatic, and relatively homogeneous. It is modelled as a 2682 m X 3166 m rectangular domain consisting of 171 rows, 158 columns, and 3 layers. Spatial discretization of the rows and columns are 30.5 m throughout most of the domain, with 7.62 m discretization in the vicinity of pumping well FLl, where a numerical pumping test was conducted. The layer thicknesses and hydraulic conductivities are spatially-variable, and were generated using sequential Gaussian simulation with parameters. The low conductivity unit in Layer 1 was assigned a hydraulic conductivity of 0.24 m/ d. The upper layer is modelled as a phreatic aquifer, and the other two layers are modelled as convertible. Specified head boundary conditions are used along the north (1403.9 m), south (1399.0 m), and east (spatially-variable constant head) boundaries. The west boundary is modelled as a no-flow boundary. The recharge rate is 2.44? lOm/d, the specific yield is 0.24, and the specific storage is 3.3? ÉÏ5 m-1. The water table map was obtained from a steady state simulation with pumping wells simulated at the model grid cells containing FLl and FL2. Steady state head values were recorded at model grid cells containing the pumping wells, test wells, and piezometers.

Well logs were obtained by recording the elevations of the top and bottom of the three model layers and the hydraulic conductivity of each of the layers at the grid cells containing each of the pumping wells and test wells. A soil type was assigned to each layer based on the hydraulic conductivity values. The water levels that are snown on the well logs are for prepumping conditions and were obtained from a steady state flow simulation with no pumping. The top of the well casing was chosen to be near the water level, and the bottom of the well casing was chosen to be at or near the bottom elevation of the Layer 3. Impermeable bedrock was assumed to exist below Layer 3.

Slug test data were obtained using the weighted arithmetic average of the hydraulic conductivity value for all layers for the model grid cells containing test wells TW2 - TVV6, where the weighting was proportional to the layer thickness. Slug test data were simulated using (2). The pumping test data were generated by running a transient simulation with pumping at the model grid cell containing FLl at a rate of 73.9 m^3/d for a five-day period. Drawdown was recorded hourly at a cell located three columns away from FLl (approximately 22.9 m east of FLl). The initial conditions for the simulation were the steady state heads from the prepumping simulation. No random error is added to the simulated drawdown values.

Evaluation

Students submitted intermediate assignments throughout the semester, as the material needed to complete the assignment was covered in class. The intermediate assignments included:

1. Sketch a water table map from the head measurements, and use it to calculate the magnitude and direction of the hydraulic gradient in the vicinity of the spill.
2. Analyse well logs to determine aquifer type (phreatic, confined) and aquifer thickness.
3. Analyse permeameter test data.
4. Analyse slug test data.
5. Analyse pumping test data.
6. Determine a representative value of hydraulic conductivity in the vicinity of the spill.

Each intermediate assignment was evaluated, and the students were provided with comments and suggestions. The intermediate assignments were graded only on effort: 10 points if the assignment was complete and reasonable effort was made to complete the assignment correctly; 5 points if only moderate effort was made to complete the assignment; and O points if the assignment was incomplete. The intermediate assignments accounted for 20% of the final project grade. The grading scheme was chosen to encourage students to make steady progress throughout the semester, and to allow ample opportunity for the students to obtain feedback.

Students prepared a final design project report which incorporated any changes that were made after receiving feedback on the intermediate assignments. The report was worth a total of 240 points (80%) of the final project grade -120 points for the correct analysis of the items in the intermediate assignments, 100 points for the design of a working hydraulic containment system and demonstration of the success of the design (with MODFLOW-2000 simulation results), and 20 points for professionalism of the final report. The students were evaluated on the accuracy of the analyses and the validity of assumptions.

Student Performance - The key parameters in designing a hydraulic containment system are the aquifer transmissivity and the magnitude and direction or the hydraulic gradient. Ideally, the injection and pumping wells should be aligned in the flow direction in order to achieve full containment; thus, the direction of the hydraulic gradient is a key parameter in the design. In addition, for a given pumping rate and separation distance between wells, the size of the contained zone depends on the aquifer transmissivity and the magnitude of the hydraulic gradient. The estimated transmissivity and magnitude and direction of the hydraulic gradient for each of the ten project teams are shown along with the values from the model that was used to create the data.

All groups except Group 6 obtained a reasonable estimate of transmissivity, with most estimates within approximately a factor of two of the true transmissivity value. The transmissivity estimates were based on the calculations of hydraulic conductivity (from the permeameter tests and slug tests) and transmissivity (from the pumping test), and on the estimate of the aquifer thickness from well logs. All available data were obtained from samples that were at least 150 m from the spill site; thus none of the data were representative of the exact conditions at the spill site. Given the sparse data, the estimates are reasonable.

All groups obtained a reasonable estimate of the magnitude of the hydraulic gradient, with most estimates within about 10% of the true magnitude of the gradient. All groups except for Group 10 obtained a reasonable estimate of the direction of the gradient. Five groups were within 9° of the true direction of the hydraulic gradient; while nine groups were within 18° of the true direction. The hydraulic gradient was estimated from a water table contour map in which the water table contours were interpolated from head data scattered throughout the region. Again, given the sparse data, the estimates are reasonable.

Contained zones for nine of the ten group designs generated these contained zones by first running a MODFLOW simulation with my base model and each group's injection well and pumping well locations and injection and pumping rates. Then, Iran a MODPATH simulation, placing particles around the injection well. Results are not shown for Group 3 because their pumping rate was not sustainable, and the cell containing the pumping well went dry in the MODFLOW simulation.

Ideally, the spill site will be completely contained inside the outermost particle paths, and all particles will be removed at the pumping well. Note that if the injection and pumping wells are not exactly aligned with the flow direction, some particles inevitably will bypass the pumping well. We consider a design to be successful if the spill site is completely contained within the outermost particle paths and if most of the particle paths end at the pumping well.

Three of the designs (Groups 1,7, and 8) successfully contain the spill site. Group 7 used a high pumping rate and a small separation distance; Group 8 used a low pumping rate and a large separation distance; while Group 1 used a moderate pumping rate and separation distance. There is no apparent correlation between the estimated transmissivity and hydraulic gradient and the success of the design. For example, Group 7 overestimated both the transmissivity and the magnitude of the hydraulic gradient, and their estimate of the direction of the gradient was off by almost 18°. Yet, their design contained the spill site because their pumping rate was sufficiently high.

Two of the designs (Groups 6 and 9) contained most of the spill site, and if the pumping rate were increased slightly, the design would be sufficient to contain the entire spill site. For Groups 2 and 5, the particle paths seem almost unaffected be the pumping and injection, indicating that the design pumping rate was too low. These groups had the two lowest pumping rates of all of the groups. Finally, the pumping and injection wells for Groups 4 and 10 were misaligned, so that the particle paths only partly passed through the spill site and none were captured at the pumping well.

The main conclusion that can be drawn from the contained zones is the need to account for uncertainty in the design parameters. If reasonable but incorrect estimates are made for transmissivity and hydraulic gradient, it is still possible to design a successful hydraulic containment system as long as the pumping rate is sufficiently high to overcome the uncertainty (e.g., the design of Group 7).

Student Perception of their Learning - Twenty-four students completed a two-part survey regarding the educational benefits of the project. In the first part of the survey, students rated nine statements on a scale of 1 to 5 about how their project experience benefited them. The statements addressed issues that are not typically covered in standard hydrogeology homework assignments. The ratings equated to :

1. Strongly Disagree
2. Disagree
3. Neutral
4. Agree
5. Strongly Disagree.

The results are shown. In all cases, the mean response was at least 4, indicating that students feel that participating in the design project enhanced their learning of hydrogeology. Several studies have shown a correlation between students' perceptions of their ability to perform technical tasks and their actual success in the performance of the task; therefore, assessment of students' perception of their learning is a valid measure of their actual learning.

In the second part of the survey, students rated six categories on whether or not the project provided a better learning experience than homework. The categories addressed activities conducted as part of the project. The ratings equated to

1. Provided a learning experience that is worse than homework
2. Provided a learning experience that is equal to homework
3. Provided a learning experience that is better than homework.

In all cases, the mean response was between 2 and 3, again indicating that students' perceptions were that participation in the design project enhanced their learning of hydrogeology. As stated above, students' perceptions have been shown to be a valid indicator of their actual performance; therefore, it can be concluded from these results that the project provided a better learning experience than traditional homework assignments.

3

The Hydrologic Cycle

"All the rivers run into the sea; yet the sea is not full; unto the place from whence the rivers come, thither they return again." Here, most succinctly, the Biblical writer has expressed what water scientists call the "hydrologic cycle." The details of how this cycle came about are not the concern of the verse writer quoted; for him it was enough to suggest the eternal circle of water. Today those men concerned with the world's water supply must look into the exact details of the hydrologic cycle.

Only when we know the nature of our resources can we intelligently plan how to provide water in plenty. That great fictional detective Sherlock Holmes once remarked that a logician could infer Niagara Falls from a single drop of water. Unfortunately, hydrology is far more complex, and logicians are far less astute, than Holmes would have us believe.

The first water was most probably not from the sea, but in the clouds. For simplicity, however, let us begin with the sea, that vast reservoir holding all but a tiny fraction of our water.

For man, to whom the withholding of a mere two quarts of the liquid a day can mean death, the quantity of water in the oceans is all but incomprehensible. To number the quarts in the sea would result in astronomical figures. Even the number of *cubic miles* contained is fantastic to those of us not used to dealing in such figures.

There are some 327,672,000 cubic miles of water in the oceans. Expressed in a way that is easier to visualise, there is sufficient water in the world that all the dry land could be covered with a column of water five miles high. Of this total, 98.33 percent is sea water. Herein lies the big problem of man dependent on fresh water to drink. Leonardo da Vinci was a pioneer in pinpointing the true nature of the hydrologic cycle. Among his many careers was that of commissioner of canals for the Milan area, and he learned enough about water to write, "we may conclude that the water goes from the rivers to the sea and from the sea to the rivers, thus constantly circulating and returning, and that all the sea and rivers have passed through the mouth of the Nile an infinite number of times." Egyptians would have been inclined to agree with this, adding perhaps that it seemed that *all* the water passed through the Nile several times during flood season. Da Vinci goes on: "The conclusion is that the saltness of the sea must proceed from the many springs of water which, as they penetrate the earth, find mines of salt, and these they dissolve in part and carry with them to the ocean and the other seas, whence the clouds, the begetters of rivers, never carry it up."

Da Vinci erred very slightly at this point, for there is some salt carried into the atmosphere. In fact, some scientists believe that salt forms nuclei for raindrops, as do ice crystals and dust particles. There is salt in rainwater, but since it is there in a concentration of only about a few parts per million, Da Vinci was essentially correct. The clouds, indeed, are the begetters of rivers. Each year about 80,000 cubic miles of water evaporates from the oceans of the world. Another 15,000 cubic miles evaporates from water on the continents. In due course, this huge amount of water must fall as some sort of precipitation. Because they represent only about one-quarter of the surface of the earth, the lands receive "only" about 24,000 cubic miles of precipitation a year.

We can thank the sun for operating this vast global distillery of fresh water for us. Aiding in circulating the vapour content of the atmosphere are the winds generated by the rotation of earth and by the uneven heating of land and water surfaces. The result is a great mixing that guarantees world tours for individual water molecules.

Sunspots, too, are thought to be connected with rainfall; the hydrologic cycle is not a simple one at all. Some idea of the tremendous power furnished by the sun lies in the fact that to provide an inch of fresh water artificially over one square mile would require the burning of 6400 tons of coal, or its equivalent, in a distillation plant.

Except in rare cases, not more than 10 percent of the moisture content of clouds is precipitated, even though it often seems that the "bottom has fallen out of the clouds." Our share of precipitation in the United States, based on the average annual rainfall of 30 inches, is about 1,564,800 billion gallons a year. This sounds sufficient for even the most wasteful nation imaginable, but much of this rainfall is lost without doing us any good. Within the hydrologic cycle is a smaller internal cycle called "evapotranspiration." Evaporation is a process we are familiar with; transpiration is the process by which plants "breathe out" moisture to the atmosphere. In humid, cool areas of the country, evapotranspiration gives back to the air about 40 percent of moisture. As we might guess, the percentage goes up in drier, warmer areas, until in the arid Southwest as much as the full 100 percent of precipitation is lost to the atmosphere. On a nationwide average, about 72 percent of precipitation is given back to the air. This means that only 28 percent is retained in or on the land for use by plants and animals. It is wrong, of course, to think that transpiration is a complete waste of water, for without it plants could not exist. However, the result is the loss of water for further use.

After evapotranspiration losses, we in the United States have about 1100 billion gallons available as "runoff" each day. This of course represents a maximum amount, and because people and plants are not always where water is, we cannot hope to be 100 percent efficient in using this "available" water. At present we are using about 320 billion gallons a day. Twice that amount would begin to strain the water budget.

The concept of runoff introduces another small cycle, or "loop," within our overall hydrologic cycle. Part of the water available after evapotranspiration losses becomes "ground water," literally water that is in the ground. This ground water takes between 33 percent and 40 percent of total runoff, and provides a natural reservoir of water for man's use if he can get to it with wells and pumps.

The ancients knew of ground water, and made use of it too, although they were uncertain as to how it came to be where it was. Ground water is not to be confused with the moisture in the soil near the surface; it is deeper down, and stored in aquifers, which are strata of permeable, or porous, rock. Hydrologists point out that underground water does not flow in "rivers" of great size, although this idea still persists in the minds of many people. Obeying natural laws, including the law of gravity, ground water seeks a lower level. Much of it comes again to the surface as springs, and it is estimated that about 40 percent of our rivers are fed in this manner.

Aquifers may be several hundred feet in thickness, and extend for many miles beneath the surface of the earth. Despite the term "water *table*," underground water is not necessarily level as is a body of surface water. For this reason, there are *artesian* wells from which water flows with no pumping. Named for the French province of Artois (Artesium in Roman times), where the first such wells were drilled in 1126, artesian water is under pressure from water at a higher level and will flow until the level drops sufficiently to reduce that pressure. Where this pressure, or "head," of water does not exist, man must lift the ground water in his well with buckets, or pump it up with a mechanical pump or one that is electrically driven.

Estimates of total ground water range from enough to cover the earth's land surfaces to a depth of 100 feet, to as much as several times that. The amount of ground water in the United States is believed to equal the amount contained in the Great Lakes. The beauty of this underground storage system is that it loses little water to evaporation, whereas the Great Lakes lose many billions of gallons each year in this way. The evaporation from the surface of Lake Mead has been measured as some 800,000 acre-feet a year. To gauge how much this is, a single acre-foot of water would provide a family of five with 180 gallons of water apiece each day for a year. The loss from reservoirs has been shown to run as high as 90 percent of the amount taken usefully from the lake.

The concept of water as a mineral, an oxide of hydrogen, is interesting with reference to the use of ground water. If a user draws from his well faster than the ground water can be replenished by precipitation of surface water, the water table drops. This is called the "mining" of water, since it depletes the supply available. Such a drop in the water table is occurring in many places in the United States and the world, although paradoxically a problem in other areas is the rising water table. This strange fact we shall discuss later; for now, let us consider only the maintained or declining amount of water in our aquifers.

Some authorities believe that ground water equals *all* our surface reservoirs and lakes, and not just the Great Lakes. The water stored in the ground is estimated to equal ten years of precipitation, or thirty-five years of runoff. This is a comforting thought, along with the knowledge that we are using only about one gallon of ground water for each four gallons of surface water for domestic, agricultural, and industrial use.

Generally, the water from wells or springs is purer than that in our lakes and rivers. It has been protected from contamination by natural causes and also from man-made pollution (with exceptions, of course). Besides this, it has been filtered through rock, sand, or other material that removes much of the contaminating matter.

An accurate description of ground water is the part of subterranean water that occurs where all the pores in the containing rock materials are saturated; that is, they hold all the water it is possible for them to hold. This zone of saturation in some cases extends all the way to the surface of the earth, as, for example, in seep areas, marshy ground, lakes, and so on. In other locations there is a layer, called the "zone of aeration," above the saturated aquifer, ranging from inches to hundreds of feet. There is water in this layer, some of it called capillary water, since it is held in the soil against the pull of gravity. But wells cannot make use of this relatively small amount of water, and Vitruvius of old was not correct in stating that moist surface soil always indicated true ground water. Moisture in the soil, while not nearly the quantity of true ground water, has still been estimated at enough to cover all the land surface of the earth to a depth of about 4 inches.

Through scientific tracing techniques, the course of underground water can be followed. Using dyes, experts have found wells being polluted from nearby sources. They have also found that water can move more than 300 miles through an aquifer. Ancient well diggers managed to dig through earth, and sometimes solid rock, to depths greater than 1000 feet. Such feats required many generations to accomplish, as we have noted. Today modern well-digging equipment can sink wells deeper than a mile, and up to five feet in diameter, in a relatively short time. When it is possible, a well is located on a high point of ground so that it can feed a large area by gravity flow, without need of an elaborate pumping system.

It is generally accepted by scientists today that ground water is replenished by the hydrologic cycle we have described; that is, by water precipitated from the atmosphere. There are, however, other theories that are interesting to consider, even though they are not widely accepted. One theory is that of subterranean condensation of water vapour from air that enters the earth. It is well known that water condenses on a cool surface, and perhaps to some extent moist air blowing on the ground adds to its water content.

Another theory holds that some ground water is "juvenile," or new, water that is constantly being manufactured from the hydrogen and oxygen that form part of the rocks far beneath the surface. Great pressures do exist at such depths, and perhaps water is formed by processes we can only guess at. Such new water is sometimes called magmatic water or volcanic water, as it is believed to be associated with magma, or molten rock from deep in the earth. Hot springs are also thought by some authorities to be at least partly juvenile or volcanic water. There is another, less controversial type known as "connate" water. This is water, much of it salty, that was trapped ages ago in folds of rock as the earth heaved and twisted during its formation. Such deposits of connate water are occasionally released by earthquakes, and add to the total water supply.

The hydrologist Oscar E. Meinzer has said, "The supply of water is not a fixed and immutable quantity, but is decreased by hydration and increased by dehydration of the rocks." Man himself adds to and detracts from the total water supply in such process as the combination of hydrogen and oxygen (in the welding torch, for example) in concrete work wherein some water is relieved of its hydrogen content and so on. But these are tiny amounts when compared with the water nature itself rearranges each year. For example, it has been estimated that plants annually reduce some 535 billion tons of carbon dioxide from water. One writer goes further, pointing out that this means the destruction of about 600 billion tons of water each year, a loss that would deplete the oceans themselves in about 21/3 million years! Since plants have been accomplishing the photosynthesis reaction far longer than 2 1/3 million years, and the seas are still here, he assumes that new water in that amount must be continually added to the "hydrosphere."

Except for traces of organic and inorganic matter, rainwater is pure. Man has added a pollutant in the form of radioactive particles from nuclear-bomb fallout, but precipitation reaching the ground is still very low in foreign matter. However, as it percolates, trickles, or flows back to the sea from whence it came, water washes salt and other minerals from the soil. This salt remains in the sea, adding to its saltiness year after year. However, ground water is endangered by the encroachment of salt water. This happens in many inland states as well as those near the sea, and steps must be taken to prevent further harming of our freshwater supplies.

Snow is part of the hydrologic cycle. This water in solid form is kept in storage on mountain watersheds, and adds to runoff during spring thaws. There are other forms of water that are important, although not so important as rain itself. These include fog, dew, and mist.

Fog is sometimes called low cloud, and if you run into clouds in the mountains you may be tempted to call them high fog. Fog occurs at ground level, and is caused by a number of factors, including humidity, temperature (both of the air and of the ground or water), air pressure, and wind velocity. In some coastal areas fog is a beneficial addition to precipitation. Tomatoes and other crops have been grown in strips of land along the California seacoast, irrigated almost entirely by fog. Whether or not forests make good use of fog is a controversial point. Some authorities claim that they do, accumulating much moisture from the fog, and using it not only for their own benefit but also for that of underbrush beneath them.

Mist is heavier than fog, and amounts to slight precipitation. Dew is something different than either. Dew contributes considerably to agriculture, and it is believed that it can be exploited far more than it has been so far. Ages ago, clever farmers irrigated plants and trees with dew by building rock cairns to condense moisture out of the air. Even on a clear day, the atmosphere contains much water, and if there were a way to extract it the desert might have a supply of water.

Dr. J. H. Dannies, a German engineer has for years experimented with equipment for extracting moisture from the atmosphere. This is done by cooling the air below its dew point. In this way Dannies has produced drinking water from the air in the Sahara Desert. More recently, Israel has investigated the possibilities of using the dew to supplement that country's meager rainfall. Tests show that in a year's time a total of 1.5 inches of water collects on the cold desert surface as dew. However, theory, and tests using methods of increasing condensation with a cold surface, indicates there should be 15 inches of dew available. Added to a normal rainfall of 5 inches, this would be a creditable total of precipitation for Israel. There are some hydrologists who claim that beneath the arid Sahara Desert in Africa is a great stratum of water-bearing rock and that all that is needed is a system of wells to tap it to make the desert once again the lush land it was described to be in ancient literature.

Floods and Drought

Water is important to life. Moreover, it must be available in the proper amounts. From the Genesis flood on, history is filled with tragic records of loss of life when rampaging water went out of control and drowned luckless human beings and animals. In 1223, for instance, Friesland in Northern Europe was flooded and 100,000 lives were lost. In 1521, Holland lost 100,000 people by drowning when the sea rose over the land. A worse flood came in 1642 when China's Hwang Ho went out of bounds and killed 300,000 people. Four years later, another 110,000 died in a flood that ravaged Holland and Friesland. In 1876, a tidal wave killed an estimated 200,000 at Bengal, India. China was struck again in 1887, and hundreds of thousands of people died when the Yellow River (Hwang Ho) flooded. Another 100,000 Chinese drowned in 1911 when the Yangtse left its banks. Another disaster involving a dam failure came as recently as 1963 when one in the Italian Alps killed 1800 people.

***Figure** : Dry Earth in the Sonoran Desert*

Such numbers of casualties make floods in our own country seem less terrible by comparison, but Americans have had their share of disaster. In 1889, a dam burst in Johnstown, Pennsylvania and over 2200 people lost their lives. Another 450 died when a dam collapsed near Los Angeles in 1928. From time to time, our country still suffers from disastrous floods. In 1951, the lower Kansas River went out of control, affecting some half a million victims. On the average, about 75,000 people are forced from their homes each year by rampaging floodwaters. Surprisingly, of the last three decades of floods that cost the United States billions of dollars plus many lives lost, the worst floods came in 1936 to 1938 when the Midwest drought was at its height. Just as too much water can be catastrophic, so can too little. One of the first recorded famines was that in Rome in 436 B.C. Ironically, thousands of the starving committed suicide by drowning themselves in the sea. Many countries were struck by the famines that occurred during the Middle Ages, in 879 and 1162. Millions have died in India, a land constantly blighted by drought, and in 1877 there were about ten million Chinese who died of starvation.

The causes of drought are not known with any certainty. It is believed that weather patterns are cyclical and that perhaps they are affected by sunspots. However, there seems no regularity in the pattern of drought. In the terrible "dust bowl" days of the 1930's, rainfall in the United States was down only about two inches from the normal average of 30 inches. Perhaps man himself is a contributor to drought and famine.

At its prime, the civilization of ancient Mesopotamia provided a good life for some 40 million people. After its fall, at the hands of invaders, less than five million inhabitants could survive. This was most likely due to the fact that the invaders destroyed the already overworked irrigation system that had made Babylon the mighty city of the ancient world. Canals filled with silt, and improperly drained fields became clogged with minerals so that they would no longer grow crops.

On the North American continent, Indians of the Southwest had a flourishing civilization built on water culture. But the Babylonian pattern was repeated, and the Pimas abandoned the lands they had laboriously watered with canals. In modern times, Arizona has repeated the cycle, adding to the salinity of the well-named Salt River Valley until it is again a problem area. In the dust bowl of the Midwest, overgrazing and erosion permitted winds to carry away precious topsoil. Not just lack of water, but the plows too ruined the land. Following World War II, parts of the United States experienced a drought that lasted a decade. The Northeast was hard hit, and the water shortage that plagued New York in 1949 focused interest on the water problem. Drought was not new, but combined with ever-increasing demand for water it created a new situation to be reckoned with.

Late in 1961, another drought began in the Northeast. By 1965 it was recognised by the Water Resources Council in Washington, D.C., as "the most intense drought in the history of that area." In a report to the President late in July, 1965, the council noted a condition of "extreme drought" in the northern half of New Jersey, the northeast quarter of Pennsylvania, the southeast half of New York, all of Vermont, all of New Hampshire, parts of Maine, and the western half of Massachusetts.

Some 300,000 square miles of area were affected by the drought. Included were heavily industrial and urban areas from Philadelphia to New York. Besides restricting the amount of water available for domestic and industrial use, the drought increased forest-fire hazards, hurt crops, and threatened further encroachment of salt water on freshwater supplies. Most alarming of all, it pointed out that there is a water crisis *now,* and not merely the danger of one in the future. Today's hydrologists are not so miraculously skilled as Sir Conan Doyle suggested. From a single drop of water they cannot infer Niagara Falls or even a smaller part of the hydrologic cycle. They are, however, much further along the road to that doubtful goal than were scientists and engineers of a few centuries ago. Man now knows the workings of nature's "waterwheel," at least in broad outline. Instead of the erroneous theories of the Greeks and the animalistic view of hydrology that Descartes advanced, water experts today possess a reasonably accurate blueprint of water in motion, as well as many bits of data both qualitative and quantitative. And this is well, of course, because as our knowledge of water has grown, so too has the water problem.

Controlling and Restoring the Waters

Water is a source of life, power, comfort, and delight, a universal symbol of purification and renewal. Like a primordial magnet, water pulls at a primitive and deeply rooted part of human nature. More than any other single element besides trees and gardens, water has the greatest potential to forge an emotional link between man and nature in the city. Water is an element of wondrous qualities. It is a liquid, a gas, or a solid. It absorbs energy and transforms it. It transports other elements in suspension and solution, shaping the landscape and nurturing life. It permeates the terrestrial environment — air, earth, and all living organisms. Pure, in the right place, and at the right time, water is an essential resource; impure, and at the wrong place and time, water is a life-threatening hazard.

An abundance of potable water is a crucial concern of all cities. To this concern, we owe some of the greatest architectural monuments of human history and some of the most impressive engineering works: the aqueducts of Rome and Nîmes and the qanÜ of Persia. Eleven aqueducts, bringing water from ten to fifty-nine miles away, supplied Imperial Rome with approximately

35 million gallons of water per day. The aqueducts delivered water to reservoirs from which it was distributed to all parts of the city. Pliny described this feat as one of the greatest achievements of Roman civilization: "But if anyone will note the abundance of water skillfully brought into the city, for public uses, for baths, for public basins, for houses, runnels, suburban gardens and villas; if you will note the high aqueducts required for maintaining proper elevation; the mountains which had to be pierced for the same reason, and the valleys it was necessary to fill up; you will conclude that the whole terrestrial orb offers nothing more marvellous."

Water availability not only determined the site of ancient cities, but also the arrangement of buildings within them. More than 3,000 years ago, the Persians first built qanâts — tunnels many miles long and up to three hundred feet deep — to carry water from mountain slopes to cities at the desert's edge. The hydraulic gradient was a measure of status. The houses and fields of the wealthy were uphill and received the water first. They used the water and passed it on. The poor, whose homes and fields were at the lowest elevations, received the water last. Stone-lined conduits, similar in design to their ancient predecessors, provide many Iranian towns with water today. The wealthy residential districts are still elevated, the poor districts depressed.

Aristotle recognised that an ample water supply was essential to both military security and health: "There should be a natural abundance of springs and fountains in the town, or, if there is a deficiency of them, great reservoirs may be established for the collection of rain water, such as will not fail when the inhabitants are cut off from the country by a war for the elements which we use most and oftenest for support of the body contribute most to health, and among those are water and air."

Urban civilizations have long grappled with the problems of water supply and use, sewage disposal, storm drainage, and flood prevention. Together, these have probably received more sustained attention throughout history than any other single urban problem. There is no lack of models for successful resolution to these problems. Urban cultures that arose in the arid and semiarid climates of Persia and the Mediterranean have developed a landscape art that both conserves and displays water. Cities like Denver, Colorado, that have reclaimed their rivers for recreation, while implementing flood prevention and water quality measures, illustrate the many social and economic benefits such projects generate. Cities that have exploited the flood storage and water treatment potential of wetlands demonstrate how parks and urban wilds can serve many uses. Most of these models, however, consist of solutions to a single aspect of the water problem: either storm drainage and flood control, sewage treatment, or water supply and conservation. The comprehensive, natural drainage system of Woodlands, Texas, a new town thirty miles north of

Houston, exemplifies the advantages of considering storm drainage, flood control, water quality, and water conservation in a single scheme. Whatever the scale from the design of a drain or a fountain to a plan for an entire metropolitan region the key to devising efficient, effective, and economical solutions is an understanding of the many ways water moves through the city.

Water in Motion

"All the rivers run into the sea, yet the sea is never full; unto the place from whence the rivers come thither they return again." The hydrologic cycle is a grand process by which rain falls on the land, is absorbed by the earth and the plants that grow in it and runs into streams and oceans, then evaporates, returning once more to the air. The power of the sun and the force of gravity drive the hydrologic cycle. The way water moves through the hydrologic cycle determines the distribution of water supplies, the occurrence of floods, and the fate of contaminants disposed of to the air, water, or land.

Only a fraction of the rain that falls on rural woods and fields runs rapidly into streams, rivers, and lakes. Leaves intercept some rain, and soil soaks up much of the remainder. Of the water that soaks into the soil, some is sucked up by plants and later returned to the atmosphere via evapo-transpiration, some evaporates directly from the soil's surface, and the remainder moves slowly through the soil as ground water. Ground water may eventually intersect the land's surface at stream beds and springs or may remain deep beneath the surface in vast underground reservoirs or aquifers. Only on steep slopes, on bare rock or ice, or when the soil is saturated, does water run off the ground's surface. The great capacity of soil and the organisms within it to absorb water and to filter and use the elements suspended or dissolved within it prevents floods, protects water quality, and conserves and restores water supplies. Traditional urban storm drainage systems short-circuit this portion of the hydrologic cycle, with disastrous results. Some cities have attempted to re-establish that link in the cycle by retaining stormwater and permitting it to infiltrate the soil; others have merely detained stormwater until the flood hazard has passed and water can be treated or safely released.

Some of the sources of water pollution — factories, sewage treatment plants, erosion from construction sites, urban runoff from storm sewers, and the fallout of dust from the air — can be pinpointed to the discharge from a specific pipe or ditch; others are more diffuse. "Point" sources are readily monitored and regulated. One can identify and measure the specific pollutants discharged, plot the precise location where they enter the water, and, given the depth and size of the water body and the circulation pattern of the water within it, predict the likely pattern of their distribution. New "point" sources, like factories or treatment plants, can be located in areas with adequate water circulation, distant from swimming beaches.

As more and more industries and municipalities conform to federal water standards, "nonpoint" sources, like air pollution and urban runoff, will become more critical water pollution problems. Nonpoint sources are extremely difficult to regulate except by collecting and treating all stormwater. Flood prevention strategies that involve the retention or detention of stormwater promise to benefit water quality, since most of the suspended solids settle out in standing water and many of the nutrients, oil, and grease are filtered out as water moves through the soil.

Storing Floodwaters

The past decade has seen a profusion of outstanding, innovative approaches to flood control by American cities. Rooftops, plazas, parking lots, and parks have been designed to store stormwater, and woods and wetlands in the headwaters preserved for their natural storage capacity, thereby reducing floods and the cost of storm drainage systems and, in some cases, permitting the treatment of urban runoff. This has generally been accomplished with little or no extra construction cost, with minimal inconvenience, and has resulted in the acquisition of new recreation land. The key to preventing floods and minimising the destruction they wreak lies in a dual strategy of storing stormwater until flooding peaks and eliminating obstacles to floodwaters within the floodplain. These principles apply whether designing a rooftop to pond and detain rain water or designating undeveloped urban wetlands as parkland to soak up and hold water in soil and plants; whether designing a pedestrian bridge so as not to block debris in floodwaters or establishing land use and building regulations in floodplains.

Floodwater storage and recreation are compatible in large urban parks. Parks that exploit the natural flood storage capacity of floodplains capture the water's edge for the public landscape. The recent profusion of urban parks that serve multiple purposes of flood control, water quality improvement, and recreation do not reflect a new idea, but rather the rediscovery of old solutions. Many nineteenth century and early-twentieth-century parks, now valued for their access to urban rivers and lakefronts, were originally designed as flood control and water quality projects.

Landscape architects and urban historians regard Boston's "Emerald Necklace" park system as a landmark in American park planning, but few appreciate that a third of the system was designed as a flood control and water quality project and not primarily for recreation.

The designer, Frederick Law Olmsted, created the Fens and the Riverway to combat the flooding and pollution problems of Boston's Back Bay tidal flats; public recreation was an incidental benefit and Olmsted himself objected to the use of the word "park" for the Fens, since he did not consider it an appropriate spot for any recreation beyond a stroll or drive along the border of the marsh. The statement printed on Olmsted's 1881 map, "General Plan for the Sanitary Improvement of the Muddy River," declares this intent:

The primary design of the scheme here shown is to abate existing nuisances, avoid threatened dangers and provide for the permanent, wholesome and seemly disposition of the drainage of Muddy River Valley. This is proposed to be accomplished chiefly by embanking, contracting and deepening the existing creek and ponds and excluding sewage and tides. The secondary design is to make use of the embankments required for the above purpose to complete the promenade here shown, of which the Common, Public Garden and Commonwealth Avenue would form about one-third already prepared and in use, and the Back Bay, now half-formed, and in progress, another third.

Until recently, historians have admired Olmsted's Boston park system chiefly for its connection of the central city with outlying suburbs, in a series of parks and connecting parkways, forgetting the flood control and water quality purpose that portions of it originally served. Olmsted designed the Fens as an irregularly shaped depression scooped out of the tidal flats. The configuration and size of the thirty-acre basin permitted the amount of water to double without raising the water level more than a few feet; during floods, twenty additional acres could be covered with water. Gently sloping banks and an irregularly shaped edge reduced waves. A tidal gate at the entry to the Charles River regulated the flow of the tides to prevent flooding and to enhance flushing of the basin. Part of Olmsted's plan was the restoration of the former salt marsh; he planted the banks of the basin with plants that could tolerate both salt and brackish water and withstand changing water levels. Olmsted felt that the juxtaposition of salt marsh and city would be novel, certainly, in laboured urban grounds, and there may be a momentary question of its dignity and appropriateness. but it is a direct development of the original conditions of the locality in adaptation to the needs of a dense community. So regarded, it will be found to be, in the artistic sense of the word, natural, and possibly to suggest a modest poetic sentiment more grateful to townweary minds than an elaborate and elegant gardenlike work would have yielded.

Portions of the Fens were planted by 1884 and within ten years had the look of a landscape that had always been there. The rapid success was largely due to the sheer quantity and diversity of vegetation planted: more than 100,000 shrubs, vines, and flowers in one area of two-and-a-half acres.

The Muddy River flows into the Fens, its current alignment and shape the nineteenth century's artificial creation. The banks of the Muddy River were regraded, lined with walkways, crossed by bridges for pedestrians and vehicles, and planted with grasses, shrubs, and trees to form the "Riverway". Like the Fens, within a few decades of construction, the Riverway had the appearance of a natural floodplain penetrating the city. Depressed below street level, with steep, wooded banks between the roadway above and the path below, it is still a retreat in the middle of modern Boston. The Muddy River survives more intact than the Fens. After the Charles River Dam was constructed in the early twentieth century, the salt marsh declined, the Fens lost the aid of the tides in enhancing water circulation, and ultimately became a dumping ground for fill from the subway and other projects.

Figure : *The Muddy River in early Spring*

Chicago, built on a flat plain only slightly higher than Lake Michigan, has been plagued by drainage and flooding problems throughout its history and has responded with ingenious solutions. In the mid 1800s, Chicago raised its street level twelve feet, jacked up and elevated existing buildings, and installed a new storm sewer system. After 12 percent of the city's population died in 1885 from cholera, typhoid, and dysentery contracted from a polluted water supply, Chicago established an autonomous regional agency, the Metropolitan Sanitary District of greater Chicago. For nearly a century, this organisation has coordinated Chicago's flood control, storm drainage, and sewage treatment. Chicago has a combined sanitary and storm sewer system and now uses stormwater detention basins located throughout the city in floodplains to detain stormwater before it reaches storm sewers, along with an extensive system of deep, underground tunnels to store the overflow from the sewer system until it can be treated. The Melvina Ditch Detention Reservoir is one of the many large detention basins operated by the Metropolitan Sanitary District and used for both flood control and recreation. Steps lead down the basin's side slope to playfields and volleyball and basketball courts in the bottom of the basin. Children ski and toboggan down the slopes of a large earthen mound at the corner of the basin and skate on an ice rink created by flooding a large, paved area near the basin's inlet. When flooded, the reservoir holds 165 acrefeet of water.

Parking lots, which account for much of the open, paved land in American cities, can also be designed to detain or even retain stormwater, as one was at the First National Bank in Boulder, Colorado, where a section of the lot can hold up to two feet of water. Consolidated Freightways in St. Louis, Missouri, constructed its parking lot to detain storm flows and netted a $35,000 savings in the cost of the storm drainage system. Outside the downtown, in less dense parts of the city, it may bc preferable to retain water long enough for it to infiltrate the soil.

Porous pavement porous asphalt, modular paving, and gravel over well-drained soils or in combination with dry wells will permit more rainfall to soak into the ground rather than run off into storm sewers. A pavement of lattice concrete blocks, with soil and grass in the interstices, is widely used in European cities, and has been employed in parts of some American cities such as Los Angeles and Dayton.

Restoring and Conserving Water

The restoration of water is also an essential function. A sewage treatment facility can be attractive and, in certain phases of its operation, compatible with recreation. In 1967, after the state of Michigan threatened to cite the city of Mt. Clemens for pollution of the Clinton River, the city combined a new sewage treatment system with a park. Combined storm and sanitary sewers comprised 90 percent of the Mt. Clemens sewer system, and sewer overflows during rainstorms had been responsible, in part, for pollution of the river. After several years' study, the city determined that collecting, storing, and treating the combined overflow was more feasible, more efficient, and less costly than separating the storm and sanitary sewer systems, and that it also offered an opportunity to create new parkland. The city constructed its new sewer overflow treatment facility with three small lakes and a park on a former sanitary landfill site. Sewer overflows remain in the first lake for anywhere from one to four days, until they can be treated in the processing building, then the water is released for aeration to the second lake for an additional seven days.

By the time the treated effluent reaches the third lake, 2.3 acres and 9 feet deep, it is appropriate for boating and fishing and for irrigating the park's landscape. In winter, when the third lake freezes, it is used for skating and ice hockey. The city plans to stock it with fish and construct a dock for summer recreation.

***Figure** : Drip irrigation system in New Mexico*

Arcata, California, exploits the properties of plants and soil to assimilate wastes, by using a wetland as part of its wastewater treatment process. Arcata renovated and reconstructed a degraded, existing wetland adjacent to its sewage treatment plant to enhance the quality of its water after treatment. The reconstructed wetland serves other functions including wildlife habitat and recreation. Other cities, including Austin, Texas, have experimented with natural or constructed wetlands to treat sewage effluents. Because wetland or aquatic plant systems to treat wastewater require more land area than conventional treatment methods, they are likely to be most appropriate for small to moderate-sized cities. The danger of introducing concentrated heavy metals and toxicants into the food chain rules out the use of such systems when effluent is heavily contaminated by these pollutants. Wetland treatment systems will be most useful in providing advanced treatment where traditional chemical methods are too costly, and they are likely to become more common as current successes become better known.

Sewage treatment can both conserve water and create an aesthetic resource. Five miles out of Santa Fe, New Mexico, a resort named The Bishop's Lodge has built a package sewage treatment facility to provide irrigation water for the resort's pasture and garden. It forms an unusual amenity in this water-poor landscape. Treated wastewater tumbles down waterfalls and cascades through sculpted channels and streams from high ground into a large pool. These "seven magic pools" provide tertiary treatment to the wastewater by aerating it and exposing it to sunlight. The water cascades a hundred feet to the resort's entrance; landscaping and earth mounds screen the treatment plant from view. Water conservation is an important benefit. Formerly, The Bishop's Lodge used 10,000 gallons of well water per day to irrigate its lawns, nearly one-third of the total daily consumption. Irrigation water now consists entirely of treated sewage effluent, an example to inspire cities to explore waste treatment that is beautiful as well as economical.

Irrigation is used routinely to maintain lawns and trees in the city, but as water shortages increase, the city must explore a more waterconserving and drought-tolerant landscape. The landscape tradition that arose in the urban civilizations of the arid and semiarid regions surrounding the Mediterranean offers many models for the modern city, for example, the protected courtyard garden or patio. The courtyards nurture lush vegetation with minimal irrigation by protecting plants from dehydrating winds and radiated heat; the barren streets of the city heighten the aesthetic relief of the courtyards. The garden art of the Mediterranean and the Middle East also exploits the many physical properties and aesthetic qualities of water with great economy. A Persian garden accomplishes a great emotional and aesthetic effect with only a trickle of water.

The subtle, refined, and profound treatment of water in the Hispano-Islamic garden makes a 100 foot jet of water elsewhere seem a vulgar display of power. An art that developed over the course of thousands of years and spread with the Moslem religion west across North Africa to Spain and east to Pakistan and India, the Islamic garden takes many forms. Each form, however, reflects the inspired manipulation of water, employing the sight and sound of water to engender a cool atmosphere of serenity and retreat. Water cascades down sculpted channels or through plain runnels into brimming basins. Slight variations in the shape of the channel produce wave patterns that catch the light in diverse ways. Water may appear precious, like a gem, as it flows over blue tiles. Water may bubble up from below the surface, or trace a graceful arc, or flow as a sheet over a molded edge. Water-poor cities should conserve their water by reserving irrigation for special or symbolic places or protected spaces where plants require minimal water. The importance of these places will thereby be heightened. Paley Park owes much of its success as an urban retreat to the contrast between its environment and the noisy, hot, dry city surrounding it. The design for Foothill College, in the semiarid climate of Los Altos, California, as originally conceived, created an oasis garden to exploit the aesthetic impact of the contrast between irrigated and nonirrigated landscape. The architects designed the college as a compound of buildings, surrounding a central courtyard, on a hilltop, with parking below.

The courtyard was designed as an oasis garden with lush vegetation sustained by irrigation; the hillside was seeded with drought-tolerant grasses. The contrast between the dry, brown hillside and the green, protected courtyard lent to the interior an atmosphere of comfort, retreat, and renewal. Since the college began to irrigate the hillside also, however, this atmosphere has been largely lost. It may be recaptured when water shortages in Northern California force the college to reduce irrigation.

In cities of a temperate, humid climate, enough rain falls to support a diverse community of plants without irrigation, so long as that water is permitted to infiltrate the soil and plants are protected from winds and radiant heat. Chestnut Park in downtown Philadelphia is paved and landscaped with plants native to that region. Rain falling within the park seeps between cracks in the pavement to the soil below. A deep layer of gravel beneath the topsoil serves both as a drainage device and as a reservoir, storing the water until plant roots can absorb it and preventing roots from becoming waterlogged. The plants have flourished and require no irrigation. Meanwhile, the park contributes no stormwater runoff to the city's sewers.

Designing to Conserve and Restore Water and to Prevent Floods

The prevention of floods and the conservation and restoration of water will only be accomplished by the cumulative effect of many individual actions throughout the city. But the impact of each will be insignificant, and might even be counterproductive, if not part of a comprehensive plan that takes into account the hydrologic system of the entire city and its region. Water pollution or flooding problems at one place may be generated somewhere else, and a solution to the water supply problem may, in the end, aggravate water pollution. The most effective, efficient, and economical solutions to urban water problems are frequently found upstream of where the problem is felt most forcefully.

***Figure** : A weir was built on the Humber River (Ontario) to prevent a recurrence of a catastrophic flood.*

The Charles River watershed is the most densely populated river basin in New England. Its headwaters are sparsely developed, but the cities of Boston and Cambridge crowd the banks of its lower basin. The U.S. Army Corps of Engineers, in a 1965 flood control study of the Charles River watershed, concluded that a new dam must be built across the mouth of the Charles River to control flooding from urban runoff in the lower basin and that over the next thirty to forty years flood-control measures upstream must be taken to prevent flooding in the lower basin. They estimated that upstream flood-control structures would cost $100 million and, instead, recommended an action requiring one-tenth the cost:

The flood control management plan recommended by this Corps' study calls for federal acquisition and perpetual protection of seventeen crucial natural valley storage areas totalling some 8,500 acres. The logic of the scheme is compelling. Nature has already provided the least cost solution to future flooding in the form of extensive wetlands which moderate extreme highs and lows in stream flow. Rather than attempt to improve on this natural protection mechanism, it is both prudent and economical to leave the hydrologic regime established over the millenia undisturbed. In the opinion of the study team, construction of any of the most likely alternatives, a 55,000 acre/foot reservoir, or extensive walls or dikes, can add nothing.

The effective role of the wetlands in flood prevention was demonstrated while the Corps of Engineers was engaged in its study. In 1968 a large storm hit Boston, and urban runoff in the lower basin crested at the old Charles River Dam within hours. The upstream peak took four days to reach the dam. The wetlands in the headwaters filled with water, gradually releasing it over the course of a month. One stretch of the river widened from fifty feet to nearly a mile. Boston's second circumferential interstate highway was under construction at the time, and because rapid urbanisation threatened the wetlands, the Corps decided that acquisition of the wetlands was the most effective method of preserving their flood storage capacity. They selected seventeen natural storage areas, ranging in size from 118 to 2,340 acres, from among 20,000 acres of wetlands in the middle and upper reaches of the Charles River. In 1974, Congress approved and appropriated $10 million to buy the wetlands for nonstructural flood control. The Corps of Engineers made the first purchase in 1977. It will retain ownership of the land, and the Massachusetts Fisheries and Wildlife Division will manage the areas as wildlife refuges.

Denver, Colorado, is an outstanding example of a city that has implemented a comprehensive, coordinated set of strategies for managing its water. The devastating property losses caused by the 1965 flood provided the incentive for the formation of the Urban Drainage and Flood Control District in 1969. Earlier, each of the region's thirty-four independent local governments had employed different methods for calculating flood risks and for designing the capacity of their storm drainage systems. Some had designed storm drainage systems to accommodate a fifty year storm; others had provided for floods from a two year storm. The Urban Drainage and Flood Control District now works with local governments to insure the adoption and implementation of adequate and consistent floodplain regulations and to undertake master plans for individual watersheds. The *Urban Storm Drainage Criteria Manual*, published in 1969, guides the work in the district and insures consistent, state-ofthe-art drainage and flood control across the entire metropolitan region. The manual covers issues of policy, law, and planning related to flood control and storm drainage, the calculation of stormwater runoff, the design of the storm drainage system, and the mitigation of flood damage.

Each year Denver's Urban Drainage and Flood Control District compiles a list for master planning of between five and ten projects for which district aid has been requested by local governments. The project must be multijurisdictional, and local governments must agree to pay half of the cost of the study and half the cost of construction, and to assume ownership after completion. The district maps the one hundred year floodplain, prepares an outline of the work to be done, and coordinates consulting engineers on behalf of the local governments. The studies cover an entire drainage basin, rather than piecemeal projects. The master plan spells out where flood problems exist and recommends remedial measures. Its recommendations will include the adoption of floodplain regulations and the implementation of such projects as stormwater detention, channel improvements, and check dams along streams to create ponds and slow stream flow. The city and county of Denver now require property owners to pay a storm drainage service charge to help finance the construction and maintenance of the stormwater system. The amount of building and paved surface on the property determines the rate billed. In 1981, when the service charge was enacted, the city estimated that annual revenues would amount to $4.7 million.

Citizens of Denver have transformed a ten mile stretch of the South Platte River, which flows through downtown Denver, from a rubble-strewn, filthy, open sewer, lined by garbage and derelict land, into a landscaped park for water sports, public gatherings, bicycling and hiking, and nature study. Like the Urban Drainage and Flood Control District, the development of Denver's South Platte "Greenway" has its roots in the disastrous flood of 1965. A flurry of investigations and reports followed the 1965 flood, but little was done about the South Platte itself until a flood in 1973, an election year, brought the issue of the river and flood hazard under the public eye again. A nine-member task force, the Platte River Development Committee, appointed by Denver's mayor and backed by over $2 million in seed money from the city, proceeded to lay plans for the river, raise additional money from public and private sources, and implement park projects.

The Platte River "Greenway" now links eighteen parks with fifteen miles of interconnected trails; with 450 acres, it is Denver's largest single park. When complete, the "Greenway" will extend twenty-five miles upstream to the foot of the Rocky Mountains and twenty miles downstream to a state recreation area on the Colorado plains.

Proponents hope that suburban communities will develop trails along the Platte's tributaries, so that eventually 120 miles of continuous river trails would lace the metropolitan region. The entire ten mile Platte River "Greenway" is now a regional centre for boating, lined by bicycle and hiking trails, and punctuated by parks. Check dams in the South Platte were designed to create white water "staircases" for canoes, kayaks, and rafts.

Competitions are now held along the man-made "Challenge Run" and slalom kayak course. At one spot, where an eight-foot dam needed to retain water for a power plant made the river impassable by boat, a boat chute was created to permit boats and rafts to negotiate the dam without portage and to serve simultaneously as a flood control device.

Carefully arranged weirs and rocks were placed to create a series of pools, riffles, and eddies ideal for recreational boating. The central channel of the Platte was excavated and large boulders and rocks placed to create a deeper stream during periods of low river flow. Water is now released from the upstream Chatfield Dam, a major flood control facility, in "recreation slugs" timed to enhance river flow for water sports during peak weekend recreation periods.

The many new parks along the Platte provide places to launch boats and to watch their progress through the chutes and slalom run. The Platte River Development Committee built the first park along the "Greenway" at the confluence of Cherry Creek and the South Platte River, where the city of Denver was originally founded. The large, terraced plaza of Confluence Park steps down to the river and provides an overview of part of the slalom run. Engineers designed the shape of the plaza and the opposite bank with a smooth profile to present minimal resistance to floodwaters, and designed the foundation to resist the river's hydrodynamic forces, by laying it directly on the riverbed and securing it with piles to the underlying shale bedrock. Years of accumulated rubble, which had blocked floodwaters and increased flood depth, were used in the construction of paths, boat chutes, and bank improvements.

An amphitheater across the river from Confluence Park was created with fill from river debris and the ruins of a bridge demolished by the 1973 flood. Pedestrian bridges, which link Confluence Park with the amphitheater and opposite banks in other parts of the "Greenway," are designed to pose no obstruction to floodwaters, since a major cause of past flood damage was the piling-up of debris at bridges in dams which diverted floodwaters into adjacent parts of the city. The wooden pedestrian bridges are designed to come loose from their concrete piers when floodwaters reach the bridge deck. Cables attached to the bridge will hold it against the downstream bank until flooding subsides. All of the parks along the floodway are designed not only to resist flood damage, but also to provide flood storage. The grading for a new bicycle path at Centennial Park, for example, was based on flood hydraulics.

With increased use of the river for walking, bicycling, and boating has come a heightened awareness of the river's water quality and a strong constituency for improving and maintaining that quality. Many sources of water pollution have been removed from the river banks as a consequence: a dump has been converted to a nature preserve; a highway maintenance yard piled with salt and sand has become Frog Hollow Park. Pressure has been brought upon the city to cease dumping street sweepings and salt-laden snow in the river. The residential neighbourhoods bordering the South Platte, several of them Denver's poorest, have gained new parks and a river environment free from former nuisances and hazards.

The Platte River "Greenway" was accomplished through the coordinated efforts of public and private organisations and individual citizens. The Platte River Greenway Foundation, established as a nonprofit, tax-exempt institution, ultimately collected over $6 million from federal, state, and local governments, from private foundations, and from individuals. The foundation, though private, cooperated closely with the city from its inception; funded and coordinated the implementation of projects on behalf of the city; and then turned over the responsibility of maintenance to the city's parks department.

Rooftops, plazas, and parking lots often provide the only space to detain stormwater in densely built downtown areas, and Denver is no exception. The city of Denver requires new and renovated buildings in the Skyline Urban Renewal District to detain stormwater on site. The alternative, upgrading the existing storm sewer system to accommodate the increased runoff, would have been prohibitively expensive and would have increased flooding in the nearby South Platte River. Developers have used a combination of rooftops, plazas, and parking lots to detain stormwater. Roofs in the Denver area are designed to support a snow load equivalent to approximately six inches of water. Engineers designed a "detention ring" to fit around the drain of a flat roof, which ponds up to three inches of water, then releases it at the rate of one-half inch per hour. A safety feature permits a severe storm to overflow the ring. Denver-area plazas and parking lots have been designed to store stormwater runoff with minimal inconvenience to pedestrians. One depressed, downtown Denver plaza, constructed above three floors of underground parking, accommodates runoff from the ten year storm; stormwater drains directly to the sewer at the rate of one inch per hour. Ponding does not disrupt use of the plaza, since elevated portions of the plaza permit pedestrians to walk across it when lower portions are flooded. Existing building codes in most American cities require that roofs be designed to withstand the equivalent of six inches of water over a short period (usually twenty-four hours), and a few cities have incorporated rooftop detention of stormwater into building codes. European cities like Stuttgart have applied the use of "wet roofs" to reduce the building's heat load as well, and thus decrease energy consumption for air conditioning. If incorporated into roof garden design, stormwater detention can also become an aesthetic amenity.

In little more than one decade, Denver has achieved considerable success in reclaiming its waters. Consider how much might be accomplished in the construction of a new city unhampered by existing buildings, streets, and drainage systems. Such a case is the new town of Woodlands, Texas, with a projected ultimate population of 150,000. When developer George Mitchell first decided to build a new town on 20,000 acres of pine-oak woodland north of Houston, he envisioned a city that would spring up in the midst of the woods, in harmony with the forces of nature. He formed the Mitchell Energy and Development Corporation and hired an interdisciplinary team of planners, engineers, scientists, and market specialists. Initially this team consisted of four firms. Over the following decade the team was expanded to include a well-staffed corporation with dozens of consultants.

By 1971, when the preliminary ecological planning study and parallel market research were complete and a general plan for the new town was underway, water had emerged as the critical factor. The Woodlands' "natural drainage system" exploits the capacity of natural, wooded floodplains to accommodate stormwater runoff and of well-drained soils to soak up and store rainfall. It reduces the combination of increased flooding and lower stream flows normally associated with urbanisation, it maintains water quality, and recharges the aquifer that underlies neighbouring Houston. The wooded floodplain, drainage channels, and recharge soils form a townwide open-space system, a natural drainage system that represents a substantial savings over the cost of constructing a conventional storm sewer system. When it was originally proposed, engineers compared the cost of the natural drainage system to that for a conventional storm system and estimated that the natural drainage system would save the developer over $14 million.

Much of the Woodlands site is very flat, with extensive areas of poorly drained soils. The construction of a traditional storm drainage system would have entailed clearance of extensive woodlands, and lowered water tables with loss of trees. It would also have increased flooding and degraded water quality downstream and, combined with an estimated 15 million gallons per day withdrawal from underlying aquifers, might have contributed to further ground subsidence under the city of Houston. The firm of Wallace McHarg Roberts and Todd, landscape architects and ecological planners, conceived a natural drainage system to resolve these problems and enable the developer to retain his vision of the future city.

The natural drainage system is composed of two subsystems: one stores and absorbs rainfall from frequent storms; the other drains floodwater from major storms. The general plan responded to the major drainage system by locating large roads and dense development on ridge lines and higher elevations, while preserving the floodplains in parks and open land, and allocating lowdensity housing to the intermediate area.

Use of the floodplains and drainage channels as open space works well from both ecological and social standpoints. Most of the spectacular trees on the site occur within the floodplains of two major creeks large, evergreen magnolias, water and willow oaks, and towering pines. These same floodplains also harbour a diverse, abundant native wildlife, including white-tailed deer, opossum, armadillos, bobcats, and many birds, and provide the corridors along which they move. The wooded easements required for drainage and flood control purposes are in most cases sufficient to insure that all but the most sensitive wildlife species may remain.

A continuous system of hiking, equestrian, and bicycle trails runs along the drainage network, linking all parts of the new town. Although this larger floodplain network drains runoff from major storms, well-drained soils and ponds absorb or store rain close to where it falls, either in private yards or in nearby parks. This local drainage system responds to subtle changes in topography and soils. Roads, golf courses, and parks are designed to impound stormwater and enhance its absorption by well-drained soils. Maintaining the structure of these soils, so essential to their ability to absorb water, required strict regulation of construction activities. Areas designated as "recharge soils" were left wooded. In some cases, building construction proceeded within a fenced-off zone that extended only a few feet on all sides from the building foundation. This practice has produced a new town with the appearance of having literally sprung up within the woods.

Models estimating the increase in peak flows generated by development at the Woodlands revealed that they will increase by only 55 percent, as compared to the 180 percent increase resulting from current "normal" development in Houston. Studies indicate that the water quality of urban runoff in the phase one portion of the new town is much better than that of other Houston residential areas. The final test of the natural drainage system occurred when a record storm hit the area in April 1979. Nine inches of rain fell within five hours, and no house within the Woodlands flooded although adjacent subdivisions were awash.

The economic benefits of a natural drainage system may not elsewhere be as dramatic as in Woodlands with its extensive flat areas and poorly drained soils, but they will accrue nonetheless. The Woodlands is and will continue to be a showpiece of drainage design, from the most mundane details of pavement and channel design to the coordination of soils, ponds, swales, and floodplains into a comprehensive drainage system.

A Plan for Every City

The successful management of water in the city will require comprehensive efforts, many individual actions, and the perception that storm drainage, flood control, water supply, water conservation, waste disposal, and sewage treatment are all facets of a much broader system. Every city should construct a framework within which both the consequences of major metropolitan efforts and the cumulative effect of individual actions can be appreciated.

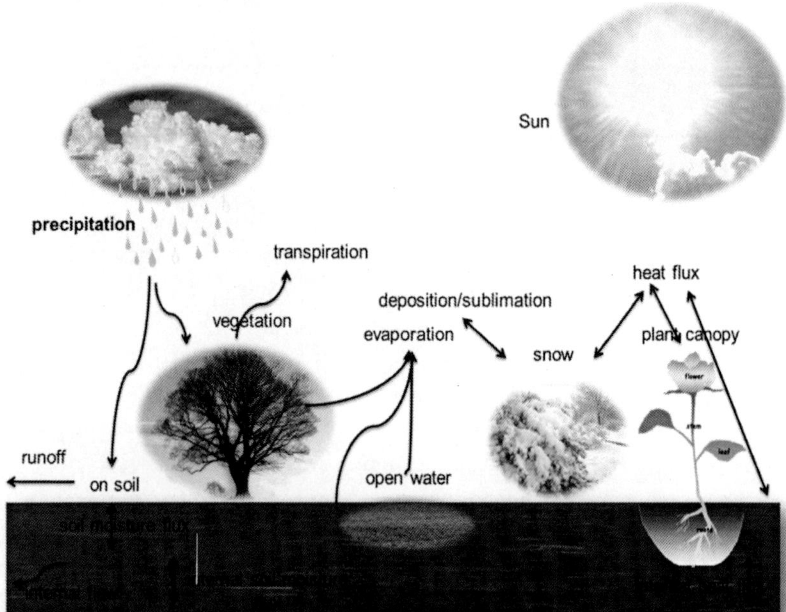

(Chen et. al., 1996, 1997; Chen and Dudhia, 2001; Ek et. al., 2003; Koren et. al., 1999)

Figure: *Many different process lead to movements and phase changes in water.*

The flow of water into and through the city including where it comes from, how and where it is used, treated, and released, and the seasonal variation of this pattern varies from city to city, depending on regional climate, topographic setting, pollution sources, and urban form. Do floods threaten a major portion of the city; and is development upstream the greater problem or constriction of the floodplain within the city? Is the city's water supply threatened by pollution of ground water or surface water, or by competing demands with other towns and cities in the region? Are large, industrial polluters the problem, or combined sanitary and storm sewage overflows? Are extremes of high flows and low flows a problem, or limited water circulation? Identifying the areas at most risk to flood hazard and those that currently provide flood storage will help in devising a comprehensive flood control strategy.

Identifying the major sources of water pollution within the city, the dispersion patterns of pollutants within surface and ground water, and water bodies with poor circulation will aid in singling out the most severely contaminated places.

Knowing the most significant water resources, those that currently provide the city with water or have the potential to do so in the future, and the areas that are most sensitive to water pollution, like aquifer recharge areas, headwater streams, lakes, and ponds, will help to preserve those resources. A comprehensive plan to prevent floods and conserve and restore the city's water should:

- address the city's most critical problems of flooding, water pollution, and water supply, with particular attention to reducing risk in the most floodprone or contaminated areas
- protect the city's most important water resources, both those currently used for water supply and those with potential to satisfy increasing demand
- locate new parks and other landscaped open space to preserve flood storage in the headwaters and the floodplain downstream, and to enhance recharge of ground water
- encourage new industry, waste disposal sites, and other polluting land uses to locate outside floodplains and ground water recharge areas that are highly vulnerable to water pollution
- locate new public buildings outside flood-prone areas and encourage new residential and commercial de velopment to do the same
- provide a plan for relocation and reconstruction after a major flood
- explore settlement patterns that would facilitate the reuse of wastewater after treatment
- exploit the flood protection and water-restoring abilities of existing wetlands
- increase the visibility of water in the city as well as public access to it.

Every new building, street, parking lot, and park within the city should be designed to prevent or mitigate flooding and to conserve and restore water resources. Every project should:

- address the relationship between the project's site and the city's critical flooding, water pollution, and water supply problems, as well as specific hazards and resources that exist on the site and in its immediate neighbourhood
- site and design buildings and landscaping to avoid flood damage
- exploit the ability of rooftops, plazas, parking lots, and the earth to detain or retain stormwater runoff design parks in floodplains to store floodwater and withstand flood damage
- design the size, depth, shape, and shoreline of urban water bodies to enhance water circulation and store stormwater
- select hardy plants that require little, if any, irrigation, fertilizers, or pesticides, and protect plants from desiccating winds
- utilise stormwater, if sufficiently uncontaminated by salts and pollutants, or treated wastewater to meet plant water needs
- exploit the aesthetic properties of water without wasting it.

Water problems and their severity vary from city to city, but every city must manage its own water resources. Cities of the past and present plagued by water problems have pioneered solutions to flood control, water conservation, and water restoration. Many of these models are applicable to every city, not just those with semiarid climates or with intensely developed floodplains. Opportunities for preventing floods, for preserving water quality, and for conserving water exist in the design of every new building and park as well as every metropolitan plan, at the centre of downtown and at the urbanising metropolitan fringe.

When the urban water crisis comes, it will probably hit fast-growing cities in and regions first, but it will extend inevitably to cities in humid regions as well. Eventually, every city will have to design a comprehensive plan for water management, including the regulation of urban form and density in headwaters and floodplains, the regulation of water use, with attendant implications for landscape design, and the careful siting of waste disposal sites and industrial and sewage outfalls.

The knowledge for such a plan exists today; the underevaluation of water is the major obstacle to its implementation. Once the water crisis forces cities to charge full value for water, the support for water conservation will follow. When cheap water is a thing of the past, rainfall will be cherished, runoff utilised, and flooding reduced. Cities will protect their water from contamination and reuse it after treatment. City parks and private grounds will acquire a drought-tolerant landscape. The use of water in public spaces will be restrained, but the impact will be powerful.

In the next decade, the dilapidated, outmoded water supply, wastewater treatment, and storm drainage systems in many older American cities will have to be overhauled. This will entail the expenditure of billions of dollars and considerable upheaval in dense urban centres. Short-term expediency must not prevail; the opportunity for redesign must be seised.

4

Groundwater Site Characterisation

In the field of management, the "Principle of Uniqueness" describes the notion that certain problems require unique solutions. It is an understanding that has resulted from unhappy experiences in attempting to apply universal solutions to the complex problems that exist in the real world. Rather than applying a structured set of prescriptive dictums or maxims to solve a recognised problem, this approach focuses on a process for achieving satisfactory outcomes. Recognition of the importance of multiple perspectives and non-linearity in real world problems has been the catalyst for the systems movement, and it applies directly to the problem of characterising groundwater flow and contaminant transport phenomena. A basic tenet of the systems movement is that there are multiple dimensions of significance (or attribution) which must be considered when evaluating a problem.

The first dimension for consideration is the level of emergence; which, according to Checkland, roughly translates into: "The system is more than the sum of its parts." The system is in some way purposeful. A system, whether human or natural, must be defined and considered as a whole entity which cannot be reduced to unconnected component activities and structures. The second dimension is hierarchy, which builds on the concept of emergence. The underlying concept is that systems are built up of smaller entities which are themselves wholes. Emergent properties (which relate to purposeful conditions such as niches or functions) denote the hierarchical level. Integral to systems thinking is the belief that a problem must be analysed at the same level as the problem itself, and not below it. This means that a problem involving a complex human or natural system must be analysed by an inclusive examination of multiple perspectives as defined by Linstone.

When considering groundwater issues, th e level or degree of site characterisation must be sufficient to identify not only the physical and chemical conditions present, but also be address the legal and sociopolitical aspects inherent in remediation (such as site history, water rights, etc.). Economics is also an important consideration; characterisation is linked directly to cost. Whether a cost is justified is based upon the regulatory climate, as well as the magnitude of the actual and perceived risk present. Perceived risk and regulatory climate are interrelated and best characterised by multiple perspectives that must be considered before site characterisation can be successfully completed.

Characterisation of Groundwater Systems

Groundwater systems are nonsteady state, dynamic and frequently subject to spatial variation. The basic physical framework is the site specific geology. Groundwater flow is dependent on both spatial and temporal factors including variations in hydraulic conductivity, gradient, porosity and thickness of stratigraphic units. Temporal variability occurs as the result of seasonal recharge, variable withdrawal of groundwater, and intermittent contaminant releases. Groundwater flow is both cyclical and noncyclical in time.

Contamination, if present, adds additional temporal and spatial uncertainty to the description of the groundwater system. As sources of contamination vary in time, the nature of the rate and extent of contamination in the groundwater changes. Chemical reactions and biotransformations also influence the kinetics of contaminant migration. For example, reductive dehalogenation is a well-documented biotic pathway for the breakdown of certain chlorinated organic substances. Abiotic pathways for chemical reaction and transformations have also been identified. Obviously, kinetic properties must be defined for the groundwater contamination problem to be properly characterised. Site characterisation is both spatial and temporal, and it requires identification of steady state and nonsteady state influences. The complexity of site characterisation and its relationship to contamination and public health concerns can be illustrated according to the procedure described by Checkland.

The hierarchy illustrates the range of issues and perspectives that must be spanned if a groundwater remediation program is to be successful in meeting the mandates set by our regulatory and legal communities. To accomplish this, characterisation of sites undergoing remediation must include an explicit approach to identify and describe the unique geological and other natural processes present, but also, to identify and accommodate what Linstone has described as the "multiple perspectives of the relevant actors" involved. He suggests that such perspectives fit into three categories: technical, organisational, and personal. These terms can be defined in the context of Checkland's systems hierarchy.

While each perspective is essential to the success of a groundwater remediation project, it is risky to assume site characterisation is complete until the technical perspective is satisfied. The problem of groundwater site remediation must be accurately defined and described at each relevant level of emergence, which at Levels 1 through 6 are principally technical in nature. Preoccupation or preference for an organisational perspective (i.e., assigning it a higher priority) can constrain and inhibit the technical perspective by restricting site characterisation resources (funds, personnel power, equipment) without regard to the technical conditions present and their implication. This short-term focus on fiscal control would therefore be applied at the expense of the risk of long-term financial exposure from liabilities associated with reduced understanding of the site. The organisational perspective, including the need for financial control, is enhanced by having a solid technical foundation. Civil engineering and the organisational response A new field of civil engineering practice has developed during the last 10 years groundwater remediation. Typical civil engineering projects involving environmental controls proceed in steps by an incremental approach.

The first engineering step is the feasibility study, which is an alternative assessment that spans the eight levels of systems emergence described earlier, including the technical and organisational perspectives of authors and intended audience. Once a project is approved for further design development, engineering proceeds in the following steps:

- conceptual design,
- preliminary design, and
- detailed design.

Engineering design occurs at the third level of systems emergence. Design is staged in a convergent, linear, programmable, and predictable sequence. The management decision process of incrementalism fits conveniently into the framework of engineering design. Applied inappropriately, this can inhibit the divergent aspects of site characterisation by restricting funding, resources, and time allocated to this important activity while prematurely emphasising the process of design.

Martin has described the "Convergent-Divergent Cycles" associated with managing technological innovation. Convergent thinking involves the inherent need to drive on to the solution of a problem, and it can involve a number of cognitive styles and tools, including morphological analysis, relevance trees, and value analysis, among others. Divergent thinking involves the generation of new ideas based on brainstorming, analogies, lateral thinking, random juxtaposition, and others. The process of bringing convergent and divergent thinking together is the fundamental task of the research and development manager such as a project manager. The process of integration involves explicit steps: problem definition, divergent thinking, convergent thinking, evaluation, and a criteria for iterating back to divergent thinking. This is a learning system.

For many years, geologists have used this approach to define and describe geological conditions at a given location. Formal utilisation of this approach brings elements of divergent thinking to bear on issues of complexity in natural processes, which is inherently part of geological field investigations. As field work progresses and information gathered, divergent thinking is used to generate as many interpretations as consistent with the data. Tests are then formulated to check the various interpretations by data already obtained or by checking against future data gathered with this in mind. Many possible scenarios can be considered, most will be abandoned and the final interpretations often bear little similarity to those postulated early in the process. The process of multiple working hypothesis may seem inefficient; it is not. It is a highly interactive learning process that has been widely utilised in other applications for many years.

It is a very cost effective technique when applied properly. It is not, however, widely taught or used by the civil engineering community. This situation is easily resolved by structuring the project team to include experienced and appropriately qualified geologists, chemists, statisticians, and managers. An important goal of the site characterisation project team is to provide the engineering team with sufficient information concerning contaminant rate and extent to design a cost effective approach to remediation. Engineering, as a formal process, begins at the feasibility study once a project has been identified. A preliminary or feasibility study is performed to consider, in detail, the implementation of alternatives and associated costs. The utility of a feasibility study related to groundwater issues is directly related to the detail and extent of site characterisation. Given a poor characterisation, no engineer can perform an adequate feasibility study, nor can any organisational perspectives be appropriately addressed. The integration of these tasks has to be one of the main objectives assigned to the project manager. Penite remediation initial engineering approach

From 1940 until 1971 sodium arsenite (Na_3~AsO_4~$12H_2O_2$~) was manufactured under the trade name of “Penite” at the former Pennwalt plant in Tacoma, Washington. Penite was used as a herbicide. It is very soluble in water and contaminated groundwater beneath four acres of the facility. In 1986, the state of Washington entered negotiations with the Pennwalt Corporation to investigate and, if appropriate, to remediate the contaminated groundwater. Consulting engineers interviewed during the selection process articulated what could be considered an organisational perspective this effort would not become “a research project.”

Divergent thinking in this case was strictly bounded for the sake of economy. When this occurs, the potential for project success diminishes significantly. Environmental investigations that began in 1987 led to the conclusion that remediation was indeed required and several options for cleanup were generated. However, these options were identified using a process managers call "bounded rationality." The concept of bounded rationality refers to the tendency of managers to select less than the best objective or alternative by engaging in a limited search for viable alternative solutions, using inadequate information, and controlling the factors influencing their decisions. Normally, the rationale for this short cutting approach to save money is supported by an underlying fear that the more we know, the more it will cost and the longer it will take. Because the scope of alternative assessment was constrained in this manner, the only apparent remediation strategies for the Penite site all involved containing the entire volume of contaminated water beneath the site forever.

Early in the investigation the engineers concluded that groundwater beneath the site simply was untreatable. This conclusion was reached after expending very little time and expertise investigating how dissolved arsenic might be treated for removal from contaminated groundwater. And, because very limited data on physical aquifer properties were collected and analysed, the consultants failed to identify a severe fatal flaw in their initial containment strategies: that an impermeable layer believed to be uniformly present beneath this site simply was not. Further, limitations in the data base forced unrealistic assumptions to be made and a high degree of uncertainty in the prediction of contaminant hydraulics. The proposed containment options ruled out any future use of four acres of prime industrial waterfront. What's more, there was no evidence or certainty that the performance objective, to reduce the net flux of arsenic in groundwater leaving the site by at least 80%, could ever be met. Corporate senior management was not satisfied with this risk.

The Penite Remediation Using a Systems Approach

In late 1988 and early 1989, Pennwalt (now Elf Atochem) reorganised the remediation project incorporating a technology research and development project management approach. The intent was to fully integrate what has been characterised as "the divergent thinking scientists (brainstorming, wild ideas, lateral thinking, etc.) and the convergent thinking typical of engineers and some managers (morphological analysis, relevance trees, value analysis, etc.)." Such assimilation of different cognitive styles was necessary to avoid the pitfalls and constraints imposed by convergent thinking prior to full consideration of possible situations and scenarios. Furthermore, since research involves a structured approach to learning, this project was structured as a learning system with information flows forward to formalize research. A systems approach was deemed more appropriate to the groundwater problem at hand because it could link together pertinent disciplines in a logical manner to create hardware, operational procedures, and subsystem linkages to accomplish a predetermined goal; and from the organisational perspective, would be very cost effective in the long term.

According to Martin, creativity is "an essential component of complex sociotechnical problem-solving." Therefore, it must be included in the management of groundwater remedial action projects. Technological perspectives include both systemic and intuitive approaches. Systematic thinking involves empiricism and rational, analytical formulations.

The successful project management approach must be able to accommodate personal perspectives, including creativity, autonomy and challenge, as well as tolerate disagreement and provide reward for successful accomplishment. This organisational perspective stimulates and sustains a learning system that integrates the diverse project team. A new project team was formed that included internal technical and managerial expertise supplemented by several outside scientists. A scope of work was formulated and additional staffing needs identified. Rather than select a single, multi-role consultant, experts were selected from various firms and integrated into a project team.

Because time constraints imposed by the regulatory community were very restrictive, several parallel tasks were essential. While additional site characterisation information was developed, a three-dimensional groundwater contamination transport model was developed. A comprehensive evaluation of hydraulic containment and control, as well as chemical engineering processes applicable to arsenic separation/removal, also was initiated. Analysis conducted during these new site investigations indicated that the containment option proposed by the initial engineering approach could not have met the site specific clean-up goals. The site had not been completely characterised and its uniqueness was not fully recognised. To understand why the containment option was inappropriate, a little site background information is required.

In the mid to late 1920s, an artificial waterway was dredged along the bed of Hylebos Creek location of the Penite site to allow marine shipping access to Tacoma's Commencement Bay in Washington State. Dredged materials consisted primarily of fine-grained silts and sands, and approached thicknesses of 8 to 12 feet beneath the existing site. Groundwater, recharged by rainfall, occurs in this deposit. The saturated thickness of the deposit is in the range of 8 to 10 feet. The former land surface was largely composed of silt and clay-sized particles with organics, such as vegetation and peat intermittently present in its thickness. This stratum is less permeable to groundwater and was described as an aquitard by the initial investigators.

This aquitard was an important component of passive containment for the arsenic contaminated groundwater. Since the aquitard was present in all 28 initial borings at the site, it was assumed to be a uniform feature. Unfortunately, the uniformity of geological features is very scale dependent. In fact, the geological processes that increased the permeability of the layer should have been postulated by geologists familiar with sedimentary facies present at this site based on a review of U.S. Geological Survey maps, U.S. Army Corps of Engineers charts and photographs from the early 1920s. Additional borings confirmed the presence of vertical migration pathways through the aquitard. Since a pathway existed between the contaminated uppermost aquifer and a deeper uncontaminated aquifer, the remedial objective of reducing the net loading of arsenic to the adjacent waterway by 80% could never have been assured by passive containment.

Kriging, a geostatistical technique employed by the mining industry for many years, was used to quantitatively evaluate site characteristics. Kriging is a statistical procedure which can be used to estimate spatial values such as hydraulic conductivity and water table elevations. It is a complex weighted averaging technique that takes into consideration the spatial relationships between data points and a grid network. Spatial data values are extrapolated at each gridnode with minimum error variance. A map of variances can be generated which reveals regions of high or low data confidence. Areas of low confidence require additional data: when the map of the site indicates high confidence, little or no additional data acquisition is warranted.

Kriging can be used as an objective, hypothesis-testing mechanism during site characterisation. It is also a useful tool for contouring spatial data. Based upon these results, site specific parameters were incorporated into the mathematical description of groundwater contamination beneath each site. This model was used to establish more refined strategies for achieving the remediation goals. It was determined that removal of concentrated arsenic in soil, reduction of onsite recharge, and placement of a steel barrier wall would reduce the discharge of arsenic in groundwater by at least 80%, thus meeting clean-up performance goals set by the regulatory community. The problem of finding treatment methods for removing arsenic dissolved in groundwater was stymied when it was discovered that high concentrations of silica were dissolved in the groundwater which resulted in a troublesome gel when conventional approaches to metals removal were attempted. Given this difficulty, the initial engineering team resorted to containment as a solution.

Decision-making by this type of process is known as satisficing. It can be problematic in certain circumstances. Interestingly, when the new project team realised that the containment option was severely flawed and turned its attention to treatment technologies, the integration of convergent and divergent thinking allowed it to look beyond the removal technologies then known by the environmental community. The team looked into creative solutions involving synergies such as separation technologies used in the chemical process industry. As a result, the project team's technology development program led to the implementation of a hybrid process for arsenic removal from groundwater. This hybrid utilises a two-step precipitation that takes advantage of a sulfide reaction and a subsequent iron precipitation step. The sulfide precipitation step has long been used for brine purification by chloralkali process engineers and precipitation by ferric chloride is currently used by environmental engineers for metals separation in Europe; the combination is, therefore, both synergistic and leveraged.

The remediation technology package was a set of integrated subsystems. The package was approved by the regulatory agencies in discrete steps, although the project team maintained an integrated perspective at all times. While this effort appeared to be incremental, no aspect of the project was ever isolated or considered "stand alone." In late 1991, the Washington State Department of Ecology approved the design of the integrated extraction and treatment system. Construction began in early 1992 and the system startup occurred in early October 1992. Options for commercialising this treatment technology are being explored. At present, the release of arsenic from the site has been greatly reduced; it is hoped that the site will be once again available for industrial use within 10 years.

Lessons Learned

Groundwater remediation projects are highly complex sociotechnical undertakings which must address multiple perspectives as defined by Linstone. Such efforts require the integration of many levels of information and the incorporation of multiple perspectives. Successful groundwater remediation is more than achieving a prescribed clean-up goal; it must be achieved with consideration of cost and long-term effectiveness. Project management approaches must accommodate more complexity than exists in most common engineering projects; site-specific factors make each effort unique. Scientific invention and engineering development must be championed and managed according to recognised needs in a supportive environment.

There are many biases that interfere with this process including a commonly encountered Fear that such projects will become endless exercises in data acquisition and cost. Yet, the tools and techniques exist and are readily available to assure that the right amount of information is obtained and the desired results achieved. Models, properly calibrated with appropriate sensitivity analysis, are important operations research tools. However, simulation models are never complete representations of reality. Models must always be suited to the problem. The cost of information, which includes simulation, must always be economically justified by the value of the decisions involved. The value of restoring contaminated property to industrial use is obvious, and it must be considered as an opportunity cost if not completed. There is value in developing new technologies which might have application elsewhere and, thus, could be marketable.

Since each site characterisation is unique, it i s evident that highly sophisticated models are not always required or appropriate. Site characterisation is itself a kind of model and simple analogues may suffice. The management concept of uniqueness applies. The real world is complex, and risk assessments and feasibility studies are often derived from limited information. While decisions should never be made mechanically based upon the output of a numerical model or statistical algorithm like kriging, models can establish invaluable range estimates when used in conjunction with appropriate sensitivity analysis. Kriging, or geostatistics, can provide highly illustrative contours of contaminant level (or other geological properties) as well as invaluable contours of the characterisation error associated with a site. Ultimately, both probabilistic and deterministic models used interactively by a qualified team incorporating multiple working hypotheses, are instrumental in effective site characterisation. The success of any program of technology development is dependent upon the effectiveness of the management process involved. Therefore, if the technical efforts associated with a remedial action are successful, the management process used must have been successful. It is all too frequent to encounter "minimalist" approaches to remediation that rely on satisfying the regulators to limit and control data acquisition. Managers of such projects seem to have forgotten that it is generally easier to regulate than to remediate.

Experience has taught that to protect corporate assets and limit liabilities, it is far better to set the scope of such activities objectively and to aggressively pursue information interactively so decisions can be made according to explicit levels of uncertainty, cost, and potential future risk. This is consistent with a long-term organisational perspective. The goal is a form of humility in decision making concerning remedial alternatives that Etzioni suggest requires decision makers to "stay loose" through the process, thus avoiding the traps of bounded rationality and incrementalism. This point is worth judicious consideration by environmental practitioners in the private and public sectors involved in groundwater remediation projects.

An Environmental Engineering Design

Educational criteria outlined by the Accreditation Board for Engineering and Technology, and supported by practicing engineers, pose a challenge to the engineering educator. Many of these criteria, such as those pertaining to problem solving and communication skills, can generally be met via the refinement of existing courses and exercises. Those concerned with enhancing problem identification and experimental design skills, functioning on an interdisciplinary team, instilling professional ethics, and promoting the desire for lifelong learning are best met by increasing the real-world content of curriculum. Infusion of effective design problems or projects into the curriculum is one way to increase the real-world content. The question then arises: How do we fit more design projects into the already overcrowded curriculum without sacrificing the depth of coverage by existing theory-based material? One answer lies in interactive simulations which are rapidly evolving as vehicles for honing realistic problem-solving skills.

An engineering design project is one type of problem-based learning (PBL) exercise. The PBL approach asks students to self-learn material in the context of a realistic problem, similar to one they would encounter as professionals in the subject area. The instructor's role in the PBL setting is to make appropriate material available to the students and provide real-time feedback in regard to the directions they choose. The motivation behind the PBL approach is that students achieve a deeper understanding when they solve realistic problems by sorting through material of variable relevance.

Examples of multimedia simulations in PBL are abundant in the medical curriculum where many of these techniques were pioneered. If a PBL exercise is to be effectively driven by such a simulation, then the underlying software must have the capacity to: 1) inspire a high level of learner activity and motivation (through subject matter depth and relevance); 2) challenge learners to make interesting and realistic choices; and 3) allow learners to make meaningful changes in the simulation parameters (i.e., let the learners experiment). The approach is well suited for the complex problems faced by medical students learning to diagnose and treat illnesses based on a combination of anatomical, physiological, and historical data. However, examples of this approach are less common in other curricula, and usually involve interactive simulations focusing on relatively specific problems, or on visualisation of complex subject matter.

In this work, we present a PBL simulator capable of delivering a broadly scoped design project. The simulator is an interactive database that administers a design project for environmental engineering students. The primary goal of this work is to advance the development of PBL simulations in the engineering curriculum. An additional objective is to infuse current contaminant fate and transport research results into an already crowded undergraduate civil engineering curriculum. An important constraint is to achieve both goals without imposing excessive time demands on the instructor. The chapter begins with an overview of software content, features and operation. The subsequent section describes the instructor-based input required by the simulator and the logistics of the design project. Finally, results from a preliminary test of the software are presented and discussed.

Interactive Site Investigation Software

The program for delivering the interactive PBL environment specific to this work is called Interactive Site Investigation Software (ISIS). In the present course, students are taught the principles of contaminant hydrogeology (physical chemistry, soil physics and groundwater hydrology) through conventional lectures and problem sets.

About halfway into the course, the students' outside assignments become virtual data collection using the interactive database. More specifically, student teams complete a design project in the form of a Remedial Investigation-Feasibility Study (PJTFS) for a virtual hazardous waste site. Virtual work tasks include drilling, collecting samples, analysing samples and performing field tests, and are designed to link theory outlined in the lectures to realworld situations.

The project discussed here was devised for a senior-level environmental engineering course. The ISIS software comprises the simulation database and graphical users' interface (GUI). The database is a set of textbased and binary files generated by the instructor or by mathematical simulations controlled by the instructor. The GUI is a Java (TM)-based (JDK(TM) 1.1, Sun Microsystems) program through which the students manipulate and query the database.

Virtual Field-work in ISIS

The design project is based on a fictitious hazardous waste site referred to as Camp Hazmat. A site plan view is used to describe the layout of this site to the students. The teams logon and start in the ISIS office by selecting a work mode. The majority of the virtual work is completed when the teams go to the field in either the soil-boring or field-testing mode. Otherwise, the teams may go to the laboratory to designate samples for various analyses, research theoretical and practical points in the virtual library, or review documents summarising the team's previously completed work These work modes are discussed further in the following sections. It is important to note that many of the choices presented to the students by: ISIS pertain to time and cost issues. When a team enters ISIS, the simulation clock is initiated and the costing algorithm begins to track all materials and services (including the team members' time).

Soil Boring: The teams begin field work by invoking the soil boring option. ISIS then prompts the team to select the drilling equipment and location (x and y coordinates). The team regulates the drilling schedule and sampling frequency in the control panel. There are three possible drilling strategies: continuous drilling with manual starting and stopping, drilling with automatic stopping at team-specified vertical increments, and continuous drilling with automatic stopping at team-specified depths or events (e.g., encountering groundwater). ISIS accelerates simulation time (16:1) during drilling and other intensive field activities in order to sustain an engaging pace. The advantages of one drilling strategy over another are time and cost related.

Figure : *Soil Boring & Pits*

With each pause in the drilling, the team has the option of designating a core sample for future laboratory analyses. There is modest cost for the time associated with sample collection and handling, and more significant costs specific to the type of analysis. Thus, budgetary constraints tend to discourage over-sampling. ISIS allows a team to save a soil sample indefinitely without repercusssions. It would be more realistic to develop a chemical-specific sample deterioration routine, and this may be considered in subsequent versions.

Once a borehole has been completed, the team must construct a well. ISIS leads the team through the process of selecting the vertical distribution of well-screening, which fixes future groundwater sampling locations. ISIS allows a maximum of three screened intervals for which the total screened length must not exceed 30% of the well. More screens allow for more vertical sampling opportunity, but the cost of well-construction increases with the number of intervals. Thus, the team must learn to differentiate between scenarios where vertical resolution of the contamination is necessary and when it is extravagant.

Field Testing: With wells in place, the field-testing option becomes viable. ISIS prompts the team to identify a well for testing and choose the desired test. Activities range from simply extracting a groundwater sample for later analysis to hydrogeologic testing (e.g., pump and tracer tests). Pump (or drawdown) tests are performed to estimate the permeability of the porous media surrounding the wellscreening. Tracer tests are performed to estimate the degree of dispersion expected during the transport of contaminants in that media. For brevity, only the single-well drawdown option is described here. Once the team chooses to execute a single-well test, ISIS prompts the team to select a pump. If the well has multiple screens, then ISIS prompts the team to select one for testing. Next, the team learns it must check the initial water level in the well and then begin pumping. Each time the team selects the "sample" button, ISIS registers the sampling time and updated water level. Water levels are calculated using analytical expressions for radial flow to a well. Calculated water levels are reported to the teams as raw data (i.e., the distance from the ground surface to the water level in the well). If at any time the estimated water level drops below the bottom of the screened interval, ISIS prompts the team to select a more appropriate pumping rate. Based on their knowledge of the well hydraulics, and on experience gained during the project, most teams learn to select reasonably optimal pumping rates so as to reduce the number of costly restarts. It is important to note that these results are based on the students' calculations (not ISIS). This approach is consistent with the current project emphasis on students learning to plan experiments and interpret results.

Analytical Laboratory: Laboratory work is subcontracted using the ISIS chain-of-custody form. Laboratory analyses currently available in ISIS include standard methods for determining chemical concentrations in soil and water samples (EPA methods) and physical soil properties (ASTM methods). Many of the standard chemical analyses screen for numerous compounds. If a given compound is not part of the instructor-generated ISIS database, then it is reported as "not detected." To submit samples to the lab, the team selects from the previously collected soil and groundwater samples, then they check the desired tests for each sample. ISIS will accept a maximum of five analyses for each sample in consideration of sample quantities needed for the analyses. Once the sample chain-of-custody form is submitted, ISIS pauses for an instructor specified amount of time (e.g., regular rate or rush), then mines the ISIS database for the appropriate analytical results. ISIS then e-mails the analytical results to every team member after the next logon.

Virtual Library: The virtual library is an optional link to HTML files containing primers on groundwater hydraulics, physical chemistry, contaminant fate and transport theory, descriptions of the drilling and sampling hardware, and standard analytical methods that the teams will be employing on the job.

Archival Documents: ISIS stores all of the work completed by each team in a series of easily accessible borehole logs, expense accounts, and laboratory reports. The teams access archival documents from within the ISIS office. The reports are especially useful near the end of the project for documenting activities and assembling the project report.

Instructor Setup Required by ISIS

Setting up the subsurface contaminant distribution in an ISIS scenario requires the creation of the ISIS database. There are four main components of ISIS database: 1) soil properties, 2) soil contaminant distributions, 3) hydraulic head distribution, and 4) groundwater contaminant distributions. The database is a set of four-dimensional arrays specifying the location and property or dependent variable value described below. The spatial discretization is the same for all properties and currently comprises a grid of 83 columns, 60 rows and 100 layers. The three-dimensional grid encompasses a subsurface volume that is about 4600 feet long by 1900 feet wide by 100 feet deep. For memory and performance considerations, the horizontal grid is irregularly spaced, and finer where the contamination is placed (where more intense probing is expected).

Soil Properties: Setup of the three-dimensional distributions of ISIS soil properties is simplified through the use of a soil palette. This palette currently contains the 15 soil types and physical properties. Thus, the instructor need only create a four-dimensional array detailing location and soil type, and ISIS will employ the soil palette to extract the relevant properties when queried about a particular location. The instructor must take care to insure that the soil type distribution is consistent with the distributed flow and transport parameters assigned to the simulation models discussed below.

Soil Contamination: ISIS tracks contaminants according to the chemical abstracts specification (CAS number). Any chemical may be used as long as its properties are provided by the instructor in the chemical properties palette. Common chemical mixtures (e.g., fuels) can also be included in the project by assigning fictitious CAS numbers and appropriate aggregated properties.

ISIS soil contaminants are distributed at various locations between the ground surface and the uppermost elevation of the groundwater domain, or water table. The soil contamination facet of the database is oversimplified in that ISIS identifies sources according to instructor-specified shape functions (cones, spheres and cylinders). ISIS only gathers contaminant data if students' queries fall within a specified shape's boundaries. Future versions of ISIS may institute a multiphase flow model to simulate the distribution of liquid spills or leaks in the unsaturated soil zone more realistically.

Hydraulic Head: The groundwater conditions are described by the hydraulic head distribution generated by a public domain groundwater flow model MODFLOW. The computational grid for the MODFLOW simulations is identical to the horizontal soil property grid, but only ten cells thick to expedite computations. Thus, the flow parameters of the MODFLOW layers must be assigned values equivalent to the average of the values for the corresponding ten intervals in the soil properties section of the ISIS database.

ISIS requires three binary output files from the MODFLOW simulations. The first file verifies the ground surface elevations input by instructor. The second reports the head distributions (i.e., equipotential surfaces) within the ten layers underlying the ISIS site. The third binary output file is necessary to drive the contaminant transport model described in the following section.

Groundwater Contamination: Contaminant distributions for the ISIS database are generated using the MT3DMS model (v. 3.50). This version allows for the simultaneous simulation of multiple solutes. The instructor must become familiar with the input structure of MT3DMS and create a set of parameters that is consistent with the assigned soil properties. MT3DMS produces a binary output file for each of the simulated contaminants. These files are required by ISIS and must be renamed according to the appropriate CAS number in order to be recognised and linked to the chemical properties palette. The distribution of the contaminant trichloroethene (TCE) within one of the MT3DMS computational layers for a portion of the Camp Hazmat site is plotted. The current ISIS database accesses ten such layers for each of the 15 contaminants.

Data Variability: An important attribute of the ISIS database is the variability that is imparted to any datum being extracted by the students. ISIS operates on each datum using a random number generator. This generator employs an instructor-specified statistical variance to synthesize a normally distributed amount of "noise" for each available physical or chemical property. The magnitude of the variance is dependent on soil and chemical type, supporting the fact that some soils and chemicals area easier to characterise than others. This aspect of ISIS was developed to exercise students' data interpretation skills. For example, students might determine that they need to repeat an observation several times before they can differentiate between a spatial trend and a measurement uncertainty.

Project Logistics: Prior to the beginning of the design project, the instructor selects the student teams and their project managers. An attempt to equalize the capabilities of the teams is made by distributing the students on the basis of their academic standing, previous coursework, performance in the present course (including 4 problem sets and a midterm examination in the current course) and relevant work experience. The students self-assign other team jobtitles (e.g., project engineer, hydrogeologist) for their group. These positions are indistinguishable in terms of work execution, which is a group effort, and are essentially placeholders in ISIS cost-allocation algorithm. The exception is the project manager, who typically a graduate student is demonstrating a comfortable level understanding of the course material to date.

In the present course offering, the instructor-developed preliminary remedial assessment (Phase I) document for Camp Hazmat is delivered during a lecture in week 5 of a 10 week quarter. This document reflects the current scenario underlying the site thereby providing students with a general idea as to the main environmental concerns in the project. The instructor summarizes the scenario in a 30 minute oral presentation, which is followed by a 15 to 20 minute discussion of how to formulate the first work plan. Later that week, each team is guided through a one hour training session outside of normal class hours. The training session prepares the students for most of the basic ISIS tasks, including borehole drilling, soil sampling, well construction and groundwater sampling. More complicated features (e.g., sample analysis, field tests) are easily learned after the basics are mastered, and are not typically needed in the early phases of the project.

Preliminary Testing

A "beta" version of the ISIS package (ISIS v. 0.9) was used to deliver the 1999 environmental engineering design project. This project had been administered manually for each of the five previous years. ISIS 0.9 lacks a costing routine and several of the hydraulic tests discussed previously. The primary objective of the beta-test was to identify problems with the code. Significant problems are highlighted throughout the following sections. A second objective was to compare the ISIS-driven project with the manually administered version in terms of design project scope and timedemands imposed (on students and instructor). Rigorous assessment of this pedagogical strategy is available for the fully functional version of the program. However, a summary of student feedback emanating from the beta-test is provided at the end of this section.

Design Project Scope

In comparison with the manually administered version of the project, the ISIS hazardous waste site encompasses a substantially more realistic problem. First, the ISIS site encompasses an area about 35 times larger than the previous site. Second, the current version of ISIS can employ any number of contaminants; at present, it includes eighteen chemicals or chemical mixtures as opposed to five that were available in the manually delivered version. Finally, the introduction of the interactive hydraulic testing modules in the ISIS-driven project represents a substantial advance from the state of the manually driven project, where a hydraulic parameter was simply another datum provided to the groups upon request.

Regarding the simulator-driven PBL environment aspect of the project, the automated data acquisition aspects of ISIS greatly enhance project relevance over the manually administered project. The manually administered project often promoted student confusion and frustration due to the instructor based lag-time associated with all data dissemination tasks. For example, a group would often request contaminant samples at certain depth intervals, only to find upon data delivery that the soil lithology had changed in the latest soil-boring location, and that the requested depth-intervals were no longer optimal in terms of sampling locations. The ISIS-driven project empowers the students to make real-time adjustments in their work plans. This aspect of ISIS alleviates student frustration due to wasted labour, and thereby helps to maintain motivation during the course of a long-term design project.

Time Demands on Students and Instructors

The teams are scheduled for two, two hour sessions per week for five weeks (20 hours total) workstation sessions in a computer lab. Each team typically held meetings for a total of about two hours per week beyond this to delegate data-reduction tasks and to plan for the next computer session. Thus, all students were committed to about six hours per week outside of lectures. In the manually administered versions of the project, students held somewhat longer group meetings to assimilate hardcopies of fresh data, but had no computer sessions.

The number of hours dedicated weekly to the design project by the instructor, teaching assistant (TA) and programmer were tracked for each the past nine years. The design project was not offered prior to 1994, and the 1993 hours represent time dedicated to problem set administration and active office hours. Manual administration of the project by the instructor (1994-1995) imposed an excessive burden. In 1996, the project was also manually administered, but most of the workload was transferred to the TA. An initial investment in programming introduced (1996) afforded some relatively minor automated data acquisition features to assist the TA. Nevertheless, project administration demands remained excessive. The demands shown for the 1999 are inflated by programming problems encountered during first two weeks of the ISIS 0.9 test. The final values shown for the years 2000 and 2001 are based on subsequent courses taught using ISIS version 1.0.

Student feedback was gathered using a short questionnaire regarding the ISIS 0.9 driven project. The primary topic in the survey pertained to the intrinsic value of the project in helping the students to understand the course concepts. Other topics included were related to workstation accessibility, time demands of the project, and the usefulness of the virtual library. The student responses were uniformly positive as to the use of the ISIS project to supplement the course lecture material. Most of the students felt that the exercise helped them to achieve a deeper and longer lasting understanding of the subject matter. All of the students were enthusiastic about the convenience of the automated data distribution via email.

There were a variety of negative responses to certain aspects of the ISIS project. One of these pertained to the time-demands imposed by the project. Approximately one-half of the students felt the project load to be excessive, or felt pressured to complete the project in an unreasonable timeframe. Several students suggested that additional academic credit be assigned to what they considered to be a laboratory component of the course. Another common source of frustration for the students was the controlled access to the ISIS workstations. Many of the students would have preferred to concentrate their ISIS-workstation endeavours earlier in the design project period instead of distributing these throughout the project. The problems with the code in the early computer sessions clearly exacerbated this issue. Transporting ISIS to the personal computer platform has been successful and will help to alleviate the problem of access. In the more distant future, an Internet-based ISIS could increase access even more substantially.

Regarding the HTML-based virtual library, students indicated that they found it helped them to understand and select appropriate analyses and tests during their work sessions. The fact that they selflearned this material is corroborated by the fact that only a minimal amount of lecture time was dedicated to these analyses. Several students suggested that the virtual library be made Internet-accessible as well as ISIS-accessible to allow them review the material outside of the work sessions. This sentiment is consistent with the notion that the limited access to ISIS during the project was frustrating to many of the students. This chapter summarizes the development and features of the Interactive Site Investigation Software (ISIS) program, a PBL environment capable of administering an engineering design project. Key findings from a beta-test of ISIS in a senior-level design project are as follows:

1) The ISIS-delivered project tested here is broader in scope and can be more easily modified from one year to the next than the corresponding manually administered project. Thus, ISIS can greatly facilitate the infusion of recent research findings into the undergraduate curriculum.
2) The automated design project relieves instructors and TAs of tedious tasks (e.g., accepting work plans and disseminating data), while allowing the learner to work at his or her own pace.
3) This automation allowed learners to work more effectively, allowing the completion of substantially more virtual fieldwork than did the manually delivered project.
4) Student's responses to the beta-test were positive with respect to perceived usability and usefulness of the ISIS program. The most common negative response addressed the restricted access to the software.

The primary intent of using a program like ISIS is to automate the tedious aspects of a design project in order to free the instructor to participate more actively in terms of providing advice and feedback to the PBL-immersed students. The improved teaching and learning dynamics will presumably produce engineers with both a deeper understanding of design process and underlying principles.

5

Groundwater in General Stream Adjudications

The federal government has water rights in all waters set aside for a federal purpose. Many states seek to delineate the size and scope of these "reserved rights" to determine what waters are available for state users. The most common way of resolving federal and state water rights claims is through general stream adjudications. General stream adjudications allow states to determine all rights to a given water source in a single lawsuit. In order to make these adjudications more efficient, Congress passed the McCarran Amendment. The McCarran Amendment waives federal sovereign immunity, enabling states to include federal water rights in general stream adjudications. In order for the waiver to take effect, however, the adjudication must be sufficiently "comprehensive." All claimants to a water source must be included in the adjudication to meet the comprehensiveness requirement. The adjudication must be more than a mere attempt by private parties to establish their water rights with respect to the federal government.

The United States may challenge a state's general stream adjudication. Specifically, the United States may move to dismiss an adjudication on the grounds that it is insufficiently comprehensive to support a waiver of federal sovereign immunity. And while the United States Supreme Court has never decided whether an adjudication must include groundwater users connected to a surface water source in order to be considered comprehensive, the Ninth Circuit Court of Appeals has answered this question in the negative. In United States v. Oregon, the Ninth Circuit held that a general stream adjudication need not include groundwater users to be comprehensive under the McCarran

Amendment. This Comment asserts that the Ninth Circuit's decision was incorrect. In addition to being incorrect as a matter of law, the practical consequences that may result argue for the inclusion of both surface and groundwater in general stream adjudications. Because states have the power to determine their own adjudication procedures, Washington should not follow Ninth Circuit precedent when adjudicating federal water rights.

Reserved Water Rights in Lands Requiring Water

Federal reserved water rights have been recognised since the early twentieth century and form the basis for numerous water rights disputes. State adjudications of water rights often implicate federal claims. The reserved rights doctrine was developed in Winters v. United States and provides the federal government with a water right at the time land is set aside for a federal purpose - if necessary to fulfill that purpose. Modern treatment of the doctrine suggests that it should be extended to groundwater rights.

Figure: *Require resorts to transfer certain types of water rights*

Tribal and Non-Tribal Lands

Most federal water rights are secured through the reserved rights doctrine, which the United States Supreme Court established in 1908 in Winters v. United States. In Winters, the Court held that the Fort Belknap Indian Reservation had an implied water right dating back to the day the reservation was established. In determining that an implied water right existed, the Court looked at the purposes of the reservation, the practical need for water, and Indian law canons of construction. The Court held that the government's purpose of moving tribes to reservations in hopes of turning them into a "pastoral" people necessarily required the recognition of water rights. Therefore, these rights did not have to be explicit in the treaty creating the reservation. The Winters decision assured tribes enough water to carry out the purposes of the reservation, with a priority date reflecting the date on which the reservation was established. The Winters doctrine allows the federal government to implicitly reserve waters when it enters into treaties with Indian tribes. Initially, the Winters doctrine was thought to apply only to Indian lands. As a result, although western states recognised the Winters doctrine, they largely ignored it.

On the other hand, the U.S. Supreme Court has addressed the merits of the Winters doctrine as well as its application to non-tribal lands. The Court's 1963 decision in Arizona v. California answered any questions about the doctrine's vitality by extending the Winters doctrine to other, non-indian, federal reservations of land. Under this framework, if Congress today reserved land for "a park, national forest, wildlife refuge, military base, or other use of public land without explicitly addressing water, the reservation of land implies Congress' intention to reserve water sufficient to accomplish congressional purposes." Similar to Indian water rights reservations in Winters, the Court established that the priority date of these non-tribal federal reserved rights is the date when the land is withdrawn from the public domain or reserved for a particular purpose.

In addition to expanding the scope of federal reserved water rights, Arizona v. California also established the "practicably irrigable acreage" (PIA) standard. This standard quantifies Indian reserved rights for agricultural purposes by looking at the irrigation capabilities of the land at a reasonable cost and allowing for a definite quantification of water rights that can be prioritised within the appropriation system. The standard troubled the western states, which contain the vast majority of Indian lands. With the large acreage of Indian lands in these states, there was a potential for extensive tribal claims to water. The PIA standard thus prompted states to begin large-scale water rights adjudications.

The Reserved Water Rights Doctrine Applies to Groundwater

The Winters doctrine refers only to surface waters, leaving uncertain whether there are federal reserved rights to groundwater. The United States Supreme Court addressed this question in 1976 in Cappaert v. United States In that case, nearby groundwater pumping by farmers decreased the water in a part of Death Valley National Monument called Devil's Hole - a deep cavern with an underground pool inhabited by a rare species of desert fish.

The Court recognised that "when the Federal Government withdraws its land from public domain and reserves it for a federal purpose, the Government, by implication, reserves appurtenant water then unappropriated to the extent needed to accomplish the purpose of the reservation." In determining a reservation of water, the Court focused on the intent of the government in setting the land aside. Intent could be inferred where unappropriated waters would be necessary to fulfill the purposes for which the reservation was made. Even though the actual water in Devil's Hole was surface water, the Court held that the United States could protect itself from damaging groundwater diversions. While Cappaert did not explicitly declare that federal reserved rights extend to groundwater, it has nevertheless served as a basis for recognising federal groundwater rights in later cases.

The U.S. Supreme Court did not address tribal groundwater rights in Cappaert, but many state courts have extended Cappaert's reasoning to these rights. The Arizona State Supreme Court was the first state supreme court to expressly acknowledge a tribal right to groundwater. In the Gila River general stream adjudication, the Arizona State Supreme Court limited the tribal right to groundwater to times when other sources were inadequate to meet the purposes of the reservation. The Montana State Supreme Court soon followed - holding that tribal reserved water rights extend to groundwater - but did not limit the right like the Arizona court. In Washington, a federal district court upheld the Lummi Indian Nation's reserved right to groundwater, also without limitation. As cases such as Cappaert and Gila River indicate, it appears that the modern trend in western states is to recognise tribal reserved rights to groundwater.

Comprehensive State Water Rights Adjudications

After federal reserved water rights were recognised in Winters, states sought to delineate water rights within their boundaries through general stream adjudications. States were not comfortable with the unknown potential of tribal water rights claims and wanted to turn hypothetical rights into quantified rights. The federal government frustrated this scheme by regularly refusing to waive its sovereign immunity. Congress attempted to remedy this problem by enacting the McCarran Amendment. The McCarran Amendment waives federal sovereign immunity in state general stream adjudications where the adjudication covers all users of a given water source.

Tribal Water Rights

In July 1952, Congress passed the McCarran Amendment, waiving federal sovereign immunity for adjudications of federal water rights. Patrick McCarran, a United States senator from Nevada, opposed the federal government's frequent refusal to litigate water rights in state courts. Senator McCarran sought to enable states to take control of their own water resources. By waiving federal sovereign immunity, his amendment paved the way for the modern general stream adjudication. States have the authority to allocate and quantify the surface and ground waters found within state boundaries. When federal reserved water rights are adjudicated in a state proceeding, state laws regarding the priority of water rights and the adjudication process apply.

***Figure** : Assert tribal water rights*

The McCarran Amendment created concurrent state and federal jurisdiction over federal water rights controversies. In Colorado River Water Conservation District v. United States, the United States Supreme Court upheld a Colorado federal district court's abstention from a tribal water rights case in favour of a concurrent state proceeding. Despite the heavy obligation on federal courts to exercise their jurisdiction, the Court pointed to multiple factors supporting the continuation of the suit in state court. These factors included the policy underlying the McCarran Amendment to prevent piecemeal adjudication of water rights.

In the same case, the U.S. Supreme Court also held that the McCarran Amendment extends to state adjudications of Indian water rights. In Colorado River, the Court noted that its previous cases concerning the McCarran Amendment did not distinguish between Indian and Non-Indian water rights. The Court additionally observed that the legislative history of the Amendment evinced a clear understanding by both its proponents and opponents that it would include tribal water rights.

Later, in Arizona v. San Carlos Apache Tribe, tribes argued that the McCarran Amendment waives only federal sovereign immunity and not tribal sovereign immunity, particularly in states that have enacted enabling statutes giving the federal government absolute control over Indian lands. The Court agreed that the Amendment waived only federal sovereign immunity, but then left tribes with the choice either to allow the federal government to uphold their water rights as a trustee or to waive tribal sovereign immunity and intervene in the state court adjudications. The Court also held that states could quantify Indian water rights in general stream adjudications despite the existence of an enabling act disclaiming state jurisdiction. However, the Court asserted that it would scrutinize any allegations of abuse of Indian water rights in state courts.

Even though the McCarran Amendment provides an avenue for state determinations of federal water rights, tribes generally prefer determination of their rights in a federal forum. As Frank Tenorio, the governor of the San Felipe Pueblo, explained, "water is the blood of our tribes."

The long history of conflict between tribal communities and the states has caused tribes to distrust the states' ability to protect tribal interests. Tensions often escalate when natural resources such as water are involved, where both the tribe and the state want access to the disputed water resource. Senator Edward Kennedy noted that "Indian water rights - no matter how critical to a tribe's future, no matter how well inventoried, no matter how brilliantly defended by government attorneys, cannot receive full protection in State court forums." Tribes thus prefer to maintain water rights proceedings in federal court whenever possible. This was made substantially more difficult following the passage of the McCarran Amendment. With firm jurisdiction over federal water rights, including tribal rights, states began to quantify these rights actively through modern general stream adjudications.

All of the Rights Asserted in the Water Source

Determining what qualifies as a general stream adjudication is one of the most contentious issues resulting from the McCarran Amendment. The McCarran Amendment itself mentions only adjudications of "a river system or other source," leaving little guidance for states to determine exactly what waters this includes. The United States Supreme Court clarified this somewhat in 1963 by establishing a comprehensiveness requirement for general stream adjudications in Dugan v. Rank. This comprehensiveness requirement means that the adjudication must be inclusive of all of the rights of owners on a stream. The suit in Dugan was a private suit to determine water rights between certain individuals and the United States. Because it was a private suit, the action was not a general adjudication of "all of the rights of various owners on a given stream" and was therefore not comprehensive enough for the McCarran Amendment to apply. By requiring that the suit include all claimants to water rights on the river and establish priority of water rights between all users, the Court established the McCarran Amendment's comprehensiveness requirement.

The U.S. Supreme Court again addressed comprehensiveness under the McCarran Amendment in United States v. District Court in and for the County of Eagle. In that case, the United States opposed an order to file a claim to rights in the Eagle River, a tributary of the Colorado River. The United States argued that Colorado's adjudication was not for a "river system" as called for under the McCarran Amendment because it did not include the entire Colorado River. The Court deemed this contention to be "almost frivolous" and held that the state need only adjudicate a river system within its own jurisdiction.

Scholars typically divide comprehensiveness inquiries into three categories: use comprehensiveness, temporal comprehensiveness, and hydrologie comprehensiveness. Use comprehensiveness focuses on how the water is used and seeks to include all uses in an adjudication. Such uses include domestic, stock watering, or agricultural uses. Temporal comprehensiveness concerns what priority dates are included in an adjudication. A temporal comprehensiveness requirement would suggest that all priority dates should be adjudicated together. Finally, hydrologie comprehensiveness focuses on the link between surface and groundwater and would require a general stream adjudication to include all hydrologically connected water sources.

Application of Hydrologic Comprehensiveness

Because the McCarran Amendment is ambiguous, it is difficult for states to decide whether groundwater must be included in general stream adjudications. Hydrologie comprehensiveness focuses on the source of water that must be included in an adjudication of federal rights under the McCarran Amendment. Scientists have recognised the hydrologie link between surface and groundwater. So has the United States Supreme Court.

Water continuously moves through the hydrologie cycle, and it is difficult to distinguish between surface and groundwater. In fact, "all groundwater in motion . . . ultimately will supply some stream. Hence, arguments that all groundwater is tributary to a stream are scientifically sound." Despite the scientific recognition of the hydrologie link between surface and groundwater, the law has been slow to acknowledge this connection. Since the first half of the twentieth century, prominent water law scholars such as Samuel Wiel have pushed states to recognise the unavoidable connection between surface and groundwater and to unify the laws regulating the two. Similarly, noted natural resources scholar Charles Wilkinson has observed that the traditional legal approach to surface and groundwater fails to acknowledge the "hydrologie reality" of their connection.

One way of responding to this "hydrologie reality" is to pursue an integrated approach to surface and groundwater. Because so much more information regarding hydrology is now available, courts are able to identify specific groundwater rights and the way in which groundwater pumping affects surface water. States adjudicating federal water rights now have the capability to map both the location of groundwater and its interaction with surface water. Some western states, such as Colorado, Idaho, Nevada, and Utah, adjudicate both surface and groundwater sources together. Even the United States Supreme Court has recognised the hydraulic link between surface and groundwater. In Cappaert v. United States, the Court allowed protection of federal water rights from nearby groundwater withdrawals because "groundwater and surface water are physically interrelated as integral parts of the hydrologie cycle."

Still, other states such as New Mexico maintain separate procedures for adjudicating groundwater and surface water claims. Likewise, in Arizona and Texas, the "hydrologie myth that groundwater is somehow separate from surface water prevails." Washington State uses the same statutory procedure to adjudicate surface and groundwater, but has yet to include groundwater in a surface water general stream adjudication.

The Mccarran Amendment does not Require Hydrologic Comprehensiveness

The Ninth Circuit Court of Appeals is the only federal appellate court to address a claim of hydrologie comprehensiveness in detail. In 1990, the State of Oregon began an adjudication of the Klamath River Basin and attempted to include the United States as a defendant on behalf of several federal agencies and as a trustee for the Klamath Tribe. The United States responded by filing suit in federal court seeking a declaratory judgment that the United States had not waived its sovereign immunity. The United States claimed that its sovereign immunity had not been waived because Oregon's adjudication was not comprehensive for purposes of the McCarran Amendment.

The United States' primary comprehensiveness argument was that the adjudication failed to include groundwater claims in the Klamath Basin. The United States stressed that the McCarran Amendment's overarching purpose was to avoid piecemeal litigation. In supporting this argument, the United States pointed to cases holding that the McCarran Amendment requires adjudicating interrelated water rights in the same proceeding. The United States argued that the McCarran Amendment's comprehensiveness requirement "makes no sense if an adjudication excludes adjudication of groundwater rights claims, the exercise of which could impair surface water rights." The United States also emphasised that the U.S. Supreme Court has never ruled on the specific question of groundwater inclusion under the McCarran Amendment, and thus no prior precedent was entirely on point.

The Ninth Circuit, in an opinion by Judge William A. Norris, rejected these arguments. The court focused first on the language of the McCarran Amendment, which states that it applies to "water of a river system or other source." The court interpreted "or" to mean that a state had a choice between adjudicating surface water or adjudicating groundwater, and held that the state did not have to do both. While the court acknowledged that one of the main purposes of the McCarran Amendment was to avoid piecemeal litigation, it relied heavily on the United States Supreme Court's decision in Eagle County. Eagle County stands for the proposition that it is not necessary to include every single hydrologically connected water source in a general stream adjudication, particularly where those water sources extend beyond the state's boundaries. The court used Eagle County to support its holding that hydrologie comprehensiveness was unnecessary and that groundwater rights need not be included.

The Ninth Circuit then analysed how the law has responded to the science of hydrologically related water sources. The court stated that the law traditionally treated surface and groundwater differently by applying different regulatory schemes to the two types of water. It also noted that different states apply riparian, absolute dominion, American reasonable use, and correlative rights doctrines to groundwater, none of which establish temporal priority by use.

In determining that some states did not apply prior appropriation to groundwater, thereby not assessing rights by temporal proximity, the court concluded that the adjudication procedure's major function - determining the priority of water rights - was absent. The court then looked to which groundwater regimes states used in 1952, the year the McCarran Amendment took effect. Because not all states used prior appropriation for groundwater in 1952, the court found that the law was not yet ready to embrace a hydrologie continuity approach to stream adjudications.

The court did acknowledge that since 1952 some states have moved toward better coordination between quantifying surface and groundwater rights. The court concluded, however, that the "recognition is too recent and too incomplete" to require disposition of groundwater rights in general stream adjudications under the McCarran Amendment. The court thus held that hydrologie comprehensiveness was not required to waive federal sovereign immunity under the McCarran Amendment.

After the Ninth Circuit's decision, the Klamath Tribe petitioned the United States Supreme Court for a writ of certiorari. The Office of the United States Solicitor General submitted a brief to the Court in opposition to the cert petition. While the Solicitor General ultimately concluded that the time was not right for the Court to review the Ninth Circuit's decision, the Solicitor General expressed the opinion that the Ninth Circuit was actually in error. The Solicitor General noted that the surface and groundwater in the Klamath Basin were hydrologically connected and that the purpose of the McCarran Amendment was to require comprehensive adjudications of entire river systems.

The Solicitor General went on to state that an adjudication excluding hydrologically connected groundwater "defeats 'Congress' purposes' by encouraging future piecemeal adjudication." However, despite the assertion that the Ninth Circuit had incorrectly ruled on the groundwater issue in United States v. Oregon, the Solicitor General recommended that the U.S. Supreme Court not grant the cert petition because the issue had not received significant consideration in other courts of appeals or the highest courts of the various states.

Hydraulic Continuity and General Stream Adjudications

Since 1944, Washington has applied the prior appropriation system to both surface and groundwater. The Washington State Legislature and the Washington State Supreme Court have both acknowledged the science of hydraulic continuity between surface and groundwater and the impact that groundwater withdrawal can have on surface waters. Washington has accepted the science of hydraulic continuity in resolving state water rights disputes. However, Washington's general stream adjudication process permits, but does not require, the inclusion of both surface and groundwater users. Even so, conflicts involving federal reserved water rights could draw from state case law to incorporate the principle of hydraulic continuity into general stream adjudications.

The Prior Appropriation System to Both Surface and Groundwater and Accepts the Science of Hydraulic Continuity

Washington State has established the prior appropriation system for both surface and groundwater. To regulate surface water, the state legislature enacted the Water Code of 1917. The Code established the prior appropriation system for surface water and created a permit system to govern surface water use. To regulate groundwater, the legislature adopted the Groundwater Code in 1945. The Groundwater Code extended the prior appropriation system to groundwater and made the permit system the sole method for obtaining groundwater rights.

Washington's statutory regime recognises hydraulic continuity between surface and groundwater. In enacting the Groundwater Code, the Washington State Legislature acknowledged the potential for conflict between surface and groundwater users and stressed the priority of surface water rights. The legislature also directed the Washington State Department of Ecology, the state agency responsible for determining water rights, to give "full recognition in the administration of water allocation and use programs to the natural interrelationships of surface and ground waters." The Washington State Supreme Court recognised this legislative intent in Rettkowski v. Department of Ecology, holding that groundwater rights should "be considered a part of the overall water appropriation scheme, subject to the paramount rule of 'first in time, first in right.'"

In addition to Rettkowski, two other Washington cases have positively addressed the hydraulic continuity between surface and ground waters. First, in Hubbard v. State, the Washington Court of Appeals upheld the Department of Ecology's decision to place conditions on groundwater permits where groundwater use would result in the Okanogan River being below minimum instream flows. The court focused on a finding of "significant hydraulic continuity" between the Wagonroad Coulee aquifer and the Okanogan River.

The court stated that where significant hydraulic continuity exists, "the groundwater rights permit must be subject to the same conditions, i.e., restrictions on withdrawal, as the affected surface water."

The second case to recognise hydraulic continuity between surface and groundwater was Postema v. Pollution Control Hearings Board. In Postema, the Washington State Supreme Court held that the state must deny a groundwater permit when there is hydraulic continuity and when groundwater withdrawal would negatively impact surface water rights. Furthermore, the court held that when a basin is closed to appropriations, a groundwater permit in hydraulic continuity with the basin must be denied if there is evidence that the withdrawal will affect flow or surface water levels. These two cases demonstrate that Washington courts recognise the necessity of addressing hydraulic continuity when permitting both surface and groundwater uses.

Statutory General Stream Adjudication Process Does Not Require Inclusion of Both Surface and Groundwater

In Washington, the general stream adjudication process is the only way to give legal certainty to water rights. General stream adjudications are governed by the Surface Water Code for surface water and the Groundwater Code for groundwater. The Washington State Department of Ecology begins the adjudication process by filing suit in superior court. Water rights holders in the adjudicated area are then notified of the suit and the holders are required to file a claim in order to become a defendant. Each claimant - the defendant in an adjudication - must file evidence with the court to support the water right they claim. The Department of Ecology then holds an evidentiary hearing regarding the water right. The adjudication statute instructs the Department of Ecology and the claimants to confer together when appropriate and also encourages settlement agreements at the hearing stage.

After the evidentiary hearing, the Department of Ecology issues a report of findings and recommendations to the superior court. Based on the evidence and findings, the Department of Ecology either files a motion for a partial decree in favour of the claims it has deemed substantiated, or a motion seeking in-court determinations of contested claims. Should a claimant disagree with the Department of Ecology, that claimant may file an exception to the report. Once all of the exceptions are decided, the superior court issues a final decree and the Department of Ecology provides a Certificate of Adjudicated Water Right for each affirmed right. This certificate represents a legally valid right and will include the priority date, purpose of use, quantity, point of diversion, place of use, and any limitations on the right. The legislature amended Washington's general stream adjudication statutes in July 2009. The changes require the Department of Ecology to review all claims filed and then move to either accept the claims or have them be determined by a court. The new legislation also encourages the Department of Ecology and claimants to "work closely together" to resolve claims outside of court. The 2009 amendments do not mandate that groundwater be included in a surface water adjudication where federal reserved water rights are at issue. The Department of Ecology, however, does have the ability to define the scope of an adjudication through both the Surface Water and Groundwater Code, with the Groundwater Code providing that the adjudication may include "either rights to the use of surface water or to the use of groundwater, or both." Thus, while the Department of Ecology can include both surface and groundwater within the same adjudication, it is not required to do so. The statute, except for fee exemptions, is silent on situations involving federal reserved water rights.

Yakima River Basin Adjudication did not Include Groundwater

The longest running general stream adjudication in Washington State is the Acquavella surface water adjudication in the Yakima River Basin. In Acquavella, the Yakima County Superior Court rejected the assertion that joinder of groundwater users is required for a waiver of federal sovereign immunity under the McCarran Amendment. In August of 1984, ten years before the Ninth Circuit's decision in United States v. Oregon, the United States filed a motion in the Acquavella adjudication to join groundwater users as necessary parties or dismiss the United States. The superior court, in a decision by Judge Walter A. Stauffacher, began its opinion by discussing groundwater use in the Yakima River Basin. The court noted that there were three principal aquifers in the region and approximately 19,000 known claimants to groundwater rights. The court also acknowledged, to the agreement of all parties, that groundwater in the area was "hydrologically connected to the Yakima River and its tributaries." The court noted that groundwater pumping would diminish surface stream flow.

The court then discussed general stream adjudications, focusing on the McCarran Amendment's text and what Congress knew in 1952 when the Amendment was passed. Because Congress had acknowledged the existence of groundwater in the Senate Report on the McCarran Amendment, the court asserted that the language "river system or other source" in the McCarran Amendment "clearly" distinguished between surface water and groundwater. The separation of "river system" and "other source" with an "or" in the McCarran Amendment was held to allow for comprehensive adjudications of either surface water or groundwater. In addition, the court found that Congress implicitly accepted Washington's adjudication process because it existed at the time Congress passed the Amendment. Finally, the court relied on the fact that it was not then "presently possible to" predict the impact groundwater pumping would have on surface water streamflow. The court maintained its jurisdiction over the United States through the McCarran Amendment and allowed the adjudication to proceed.

Groundwater in General Stream Adjudications Involving Federal Water Rights

Washington should require the inclusion of hydrologically connected surface and groundwater users when adjudicating federal reserved water rights. The Washington State Supreme Court should reject the Ninth Circuit's reasoning in United States v. Oregon. The Washington State Supreme Court is not bound by Ninth Circuit precedent and can choose to interpret the McCarran Amendment to include groundwater. Washington should also re-examine the Yakima County Superior Court's outdated understanding of hydraulic continuity in the Acquavella adjudication. By doing so, the Washington State Supreme Court would re-affirm the connection between surface and groundwater and protect federal reserved water rights, including the important water rights of Indian tribes. Adjudicating surface and groundwater users together would also avoid piecemeal litigation, accomplishing the McCarran Amendment's primary purpose. This is especially true as Washington faces ongoing water rights disputes across the state in the years ahead.

The Ninth Circuit's Flawed Reasoning

The Ninth Circuit's decision in United States v. Oregon that groundwater was not necessary for a comprehensive adjudication under the McCarran Amendment was incorrect for two reasons. First, the Ninth Circuit failed to distinguish groundwater users from the out of state water right holders at issue in Eagle County. This omission ignores the importance of hydrologie continuity, similar to the Yakima County Superior Court's ruling in the Acquavella adjudication. Second, the Ninth Circuit incorrectly reasoned that water law treats surface and groundwater separately. This narrow conception of water law ignores important considerations for treating surface and groundwater together, including the consistent adjudication of tribal water rights.

The Ninth Circuit placed too much precedential value upon the United States Supreme Court's decision in Eagle County. In Eagle County, the United States argued that all connected water sources need to be adjudicated at the same time in order to join the federal government under the McCarran Amendment. The Court properly held that the McCarran Amendment's comprehensiveness requirement did not require states to include all connected water sources that ran into other states. In its heavy reliance on Eagle County, the Ninth Circuit failed to properly distinguish in state groundwater users from the out of state water users at issue in Eagle County. In Eagle County, the U.S. Supreme Court said the government's argument for state adjudication of out-of-state water rights was "almost frivolous." Arguing that hydrologically connected surface and groundwater be adjudicated at the same time is not "almost frivolous." It is an argument based on sound science and a growing legal trend. The U.S. Supreme Court itself has acknowledged that "groundwater and surface water are physically interrelated as integral parts of the hydrologie cycle." The unfounded claims rejected in Eagle County should not tarnish the important and legitimate connection between surface and groundwater.

In Washington, the Yakima County Superior Court made a similar ruling to United States v. Oregon, holding in 1985 that a general stream adjudication did not require groundwater. This decision should not be conclusive. First, the court recognised hydraulic continuity, a science that has been further developed and incorporated into the Washington legal system in cases such as Hubbard and Postema. Importantly, in Postema, the Washington State Supreme Court stated that "The Department of Ecology may use new information and scientific methodology as it becomes available and scientifically acceptable for determining hydraulic continuity and the effect of groundwater withdrawals on surface waters." The inclusion of both surface and groundwater in future adjudications would be consistent with this important directive.

In addition to its improper reading of Eagle County, the Ninth Circuit in United States v. Oregon incorrectly reasoned that state water law treats surface and groundwater separately. The court quoted a portion of Beck's treatise Waters and Water Rights indicating that states have not adopted a legal regime for uniform treatment of surface and groundwater. By relying on this narrow statement, the court failed to acknowledge states, such as Washington, that do recognise the science of hydrologie continuity and have integrated it into water rights adjudications. Simply because states have not adopted a uniform system for managing surface and groundwater together does not mean states should not adjudicate surface and groundwater together. There are circumstances in which the two must be considered together, such as when a state is attempting to quantify tribal water rights.

Furthermore, the United States Solicitor General's brief in opposition to the Klamath Tribe's petition for writ of certiorari supports a finding that the Ninth Circuit was in error. The Solicitor General explicitly stated that the Ninth Circuit was wrong in ruling that the McCarran Amendment does not require the inclusion of hydrologically connected groundwater. The Solicitor General referenced the congressional intent behind the Amendment of the need to avoid piecemeal litigation of water rights and emphasised that comprehensiveness should include hydrologically connected groundwater. Although the cert petition was denied for lack of extensive consideration by other courts, perhaps now, nearly seventeen years after the Ninth Circuit's decision, the time is ripe to re-address this issue.

In Washington, there is even another reason not to follow the Ninth Circuit: the importance of water to tribal communities and the hesitancy of tribes to trust state court adjudications of water rights. These factors argue in favour of including groundwater in general stream adjudications. Even though the United States Supreme Court promised exacting scrutiny should a state court abuse tribal water rights, adjudications can "linger indefinitely in the lower courts," waiting for the final judgment that is needed for appeal to the U.S. Supreme Court. State court adjudications of tribal water rights should therefore be allowed to proceed only where all the prerequisites for a waiver of federal sovereign immunity under the McCarran Amendment are met, including hydrological comprehensiveness.

Including Groundwater in General Stream Adjudications Comports

In passing the McCarran Amendment, Congress sought to prevent piecemeal litigation of federal reserved water rights. Subsequent U.S. Supreme Court cases have recognised Congress' intent. Requiring that surface and groundwater be adjudicated together would be consistent with Congress' intent to promote holistic, as opposed to piecemeal, litigation. General stream adjudications including both surface and groundwater rights will allow the federal government to litigate its water rights only once for each water source. Conversely, treating surface and groundwater separately would require the federal government to participate in multiple legal actions, with the possibility of decades lapsing between adjudications due to their high cost and extensive scope.

In Washington, the Yakima County Superior Court rejected hydrological comprehensiveness in the Acquavella adjudication based on the court's interpretation of the McCarran Amendment. This interpretation, however, ignores Congress' intent in passing the McCarran Amendment - to avoid piecemeal adjudications of federal water rights. It is only through comprehensive adjudication of both surface and groundwater that priorities and quantities of use in a hydrologically connected water source can be articulated with sufficient legal clarity. This is necessary to enable holders of federal reserved water rights to enjoin interfering uses. As the Acquavella adjudication in Washington draws to a close after more than thirty years, a second adjudication for groundwater rights in the Yakima River Basin may be necessary. This further demonstrates the inefficiency of not adjudicating groundwater rights during the surface water adjudication. Before that proceeds, the Department of Ecology will likely begin an adjudication of the Spokane River area. Any future general stream adjudication in Washington that includes federal reserved water rights should address both surface and groundwater use. This will guarantee compliance with the McCarran Amendment's comprehensiveness requirement and ensure waiver of federal sovereign immunity for the adjudication.

If a state wishes to adjudicate a federal water right in a general stream adjudication, then federal sovereign immunity must be waived under the McCarran Amendment. The United States Supreme Court has determined that the waiver of sovereign immunity is effective only when the adjudication is comprehensive, meaning that it covers all of the rights in the river system being adjudicated. The Court's comprehensiveness requirement for the McCarran Amendment comports with Congress' intent to avoid piecemeal adjudications of federal water rights. In addition, the tribal preference for a federal forum in water rights determinations further stresses the importance of only waiving sovereign immunity when a state adjudication is truly comprehensive. Scientists recognise the hydrologie connection between surface and groundwater, and the law has increasingly come to acknowledge this link. A comprehensive adjudication under the McCarran Amendment should always include hydrologically connected surface and ground waters. Without a requirement of hydrological comprehensiveness, a surface water adjudication would be incomplete. Numerous groundwater users would be left out of the adjudication and would not have clearly defined rights in respect to their connected surface water counterparts. This would result in the federal government, including tribes, having to litigate for groundwater rights separately.

Will increased demand for groundwater and new population pressures put western farms out of business? Most people take water for granted. As long as it flows out of the tap in a relatively unadulterated form and doesn't look, smell, or taste bad, we give very little thought to where water comes from, how it got into our homes, or what impact changing water uses will have on our communities. In the arid western United States, however, one of the greatest natural-resource management problems in coming decades will be the quantity of available groundwater, its quality, and how it is to be used.

In large sections of the West, groundwater is the only dependable source of water. For the 19 western states as a whole, roughly 35 percent of the water consumed comes from the ground. Furthermore, dependence upon groundwater in many areas will increase in the future since the best surface water sources rivers, lakes, and streams have largely been developed or appropriated. As dependence on groundwater increases, a number of subtle shifts in local management policy, economics, and uses will occur. In areas where inexpensive surface water has been available, for example, additional agricultural expansion will be possible only through pumping more-expensive groundwater. In addition, population shifts from rural areas to cities will cause changes in patterns of land use, economic activity, and the sociopolitical composition of the area. The twin pressures of existing agricultural water use and the need to facilitate new population concentrations will, without proper management, eventually deplete groundwater supplies through overdrafting, as water is extracted at a rate that exceeds recharge. Moreover, as groundwater supplies become depleted, and the cost of water increases, the rising price will result in changes in water usage. In some areas, groundwater will become uneconomical for agricultural use, and the lands will revert to dry farming or go out of production altogether.

These changes have social, political, aesthetic, and economic impacts and will change the character of local communities and the West as a whole. The importance of groundwater cannot be overstated.

Resource at Risk

It is estimated that more than 90 percent of freshwater in the United States is in the form of groundwater. The balance is in lakes, rivers, and streams. Use of groundwater has increased dramatically in the recent past. From 1945 until 1980, groundwater use more than quadrupled, increasing from 21 billion to 88 billion gallons a day. In the mid-1990s, approximately one-half of all the people in the United States used groundwater as a primary source of drinking water. Nationally, groundwater provides 39 percent of the water used in public supply systems. Groundwater also provides 75 percent of the water used for rural domestic consumption and livestock purposes, 37 percent for irrigated agriculture, and 2 percent for self-supplied industrial purposes. In the West, however, where 33 percent of the freshwater comes from groundwater sources, most of the water, 78 percent, goes for irrigated agriculture.

Public supply systems consume 14 percent, and rural domestic and livestock use accounts for 4 percent. Although these percentages vary significantly from state to state, it is clear that in the West, where agriculture plays a large role in the economy, groundwater will weigh heavily in determining the future and direction of the economy. And, as we might expect, shifts away from agricultural uses will have ripple effects throughout those economies. Unfortunately, despite its obvious importance, many people do not understand the nature of groundwater.

The Resource

When precipitation falls to the earth as rain or snow, some percolates through the soil to a water table, which is the top of what is referred to as the zone of saturation. A zone of saturation is also fed by surface water from wetlands, lakes, and streams. Water that reaches the zone of saturation from the surface is said to percolate; hence, such water is sometimes referred to as percolating groundwater. Whatever the source, groundwater moves very slowly, its actual speed depending on the composition of the material in which it is held. When the water in a saturated zone is sufficient and the composition of the material is adequate to store amounts of water for human use, it is referred to as an aquifer.

There are two types of aquifers: confined and unconfined. Confined aquifers, also called artesian aquifers, are overlaid by impervious matter such as clay or shale. These aquifers are under greater than atmospheric pressure; hence, the water sometimes flows to the surface. Unconfined aquifers are those that are under atmospheric pressure, and water level variability in such aquifers depends on natural conditions, particularly the rate of replenishment, or recharge, to the water table. When water is withdrawn from an aquifer at a rate that is greater than its rate of natural or artificial replenishment, the aquifer is said to be mined, or overdrafted. When such conditions continue indefinitely, the water table drops and the resource is depleted. Then, of course, pumping stops. In such situations pumping often stops before total depletion because the increased cost of drawing the water from greater depths makes further use of the water uneconomical.

Evolution of Technology

Although humans have been taking water out of the ground since ancient times, it has only been within the last century indeed within the last few decades that the means have been developed to withdraw large amounts of water from the ground. For most of the history of the West, the major problem has been how to get water out of the ground. The first successful and widely adapted device for extracting groundwater was the windmill. Although windmills took various shapes and forms, the early designs all had in common a dependence on erratic wind patterns, inefficiency, and overall low productivity. A windmill with a 25-foot diametre wheel, which was large, operating in a 16 mile/hour wind was capable of lifting approximately 37 gallons of water a minute from a maximum depth of 70 to 80 feet. Although such yields were sufficient for domestic uses, they did not allow for widespread irrigation.

After the windmill came the centrifugal pump. Early centrifugal pumps had a number of serious drawbacks. First, they had to be located within 20 feet of the water table. This sometimes required quite deep and relatively wide pits, with the pump operating at the bottom of the pit. Since these pumps were expensive to purchase and operate, their use was limited to high-value crops such as rice. These early pumps were also often belt driven and powered by steam, requiring frequent adjustments and constant care. Later, steam was replaced by low-compression oil-burning engines. The cost of such pumps in the first decade of the century was approximately $4,000.

Through the 1920s, pump designs improved. Centrifugal pumps were developed that allowed for operation at much greater depths and with higher capacity. Inefficient steam and low-compression oil-burning power plants were replaced by hooking up an old car motor on a direct drive to the pump. By the 1920s, the cost of the pump, power plant, and drilling had dropped to approximately $2,000. Further improvements in pump design in the 1930s and the availability in many areas of cheap electric power combined to rapidly increase the number of wells and the yield of these wells throughout the West and the Midwest. By the end of the decade, the technological problems associated with groundwater pumping had largely been overcome, and concerns over water supply began to emerge.

Improvements in pump design allowed the total amount of groundwater withdrawn for western irrigation to increase from relatively small amounts in the early 1930s to about 10.7 million acre-feet by 1945 and up to 56 million acre-feet by 1975. Most of these early expansions took place on the high plains, a semi-arid region characterised by relatively low rainfall averaging less than 20 inches, or about 50 centimetres and climatic extremes. From World War II to the mid 1960s, the greatest increases in groundwater use were concentrated in the Southern Plains, most notably in the area surrounding the Texas Panhandle. Groundwater withdrawal for irrigation in the Southern Plains grew from less than 1 million acre-feet in 1945 to approximately 13 million acre-feet by 1965. The same technological advancements allowed for increased use of groundwater in the Northern Plains, home to 28 percent of the nation's farmland, although that expansion occurred somewhat later. In 1950, 26 percent of the water used in irrigation in the Northern Plains came from the ground. By 1975 this had increased to 78 percent.

The development of centre-pivot sprinkler irrigation systems has also facilitated the expansion of groundwater use in the West. First patented in 1952, centre-pivot irrigation systems are made up of a long line of sprinklers that move in a circle around a fixed pivot like the large hand of a clock. Water is supplied from the central pivot point. Most systems are designed to irrigate a quarter of a section of land 160 acres but because of the circular pattern, they actually irrigate only 130 acres. Sprinkler irrigation and centre-pivot irrigation have the advantages of using less water than flood or furrow irrigation. Sprinklers may be used in areas with sandier soil or hillier land than is possible with other types of irrigation. Also, given that one person can irrigate from 10 to 15 quarter sections 1,300 to 2,000 acres-centre-pivot systems can reduce labour costs significantly.

Evolution of the Law

The technological changes that facilitated rapid expansion of groundwater pumping and the development of irrigated agriculture in the West put pressure on the legal doctrines governing groundwater use and ownership. As it became possible to withdraw greater amounts of water from greater depths, competition for water developed in some areas. Legislatures in the West reacted to conflict over groundwater resources in a variety of ways. They changed their laws or perhaps not depending upon the controversies involved the participants, the interests, and the pressure lawmakers felt. Because of early conflicts over groundwater, some states New Mexico, for example were quick to write relatively comprehensive groundwater management statutes. In other states, notably Texas and California, early water law has changed greatly yet retained significant parts of the law as it existed prior to the rapid development of groundwater resources.

In some states, the line between the old and the new laws is not always clear, and this lack of clarity amidst competition for and demand on the resource has sometimes led to what many feel are inefficient uses and overutilisation of groundwater.

There are, in fact, four major groundwater law doctrines in the United Stares, and groundwater law has changed since the turn of the century.

The four primary legal doctrines governing groundwater use are the English or common law rule of absolute ownership, the American rule of reasonable use, the correlative rights doctrine, and the doctrine of prior appropriation. Generally, groundwater law in the western states has evolved during this century from the English, or common law, rule of absolute ownership to either the American rule of reasonable use or, in most western states, the doctrine of prior appropriation.

Absolute Ownership

The common law, or absolute ownership doctrine, holds that the water beneath one's land is the property of the landowner and may be withdrawn, without malice, with no regard to the effect that withdrawals have on any other landowner. In theory, and in practice in many areas, this has meant that landowners could pump at will the water beneath their lands as well as the water beneath the lands of their neighbours. The absolute ownership doctrine was developed in England and transferred to the relatively wet East, where it is largely still the law.

This doctrine works reasonably well in areas where water is abundant. However, in the more arid western states, many courts and legislatures, familiar with water law in other parts of the country, have adopted the common law rule.

In early western history, minor amounts of groundwater were withdrawn because of technical obstacles and relatively small demand. Since competition for groundwater resources had not yet become an issue, the absolute ownership doctrine seemed the reasonable course to pursue. However, when competition for water did develop in the West, the drawbacks to the absolute ownership doctrine in an arid environment became apparent. It was shortly after competition for water developed as early as the late 1800s that modifications of the rule started to be made.

One case that dealt with problems in the rule was Bassett v. Salisbury Mfg. Co. (1862). In this case, the court recognised the potential dangers of absolute ownership for both the resource and for users without high capacity pumping devices. Short of any modification, it was clear that there was nothing in place to dissuade high-capacity pumpers from depleting the resource and leaving other users at their mercy. Moreover, prior to any change in the rule, even malicious depletion aimed at intentionally harming a neighbour was legal under strict interpretation of the law.

American Rule

One modification made by many courts in the West was the reasonable use doctrine, or the American rule. Basically, the reasonable use doctrine limits a landowner's right to the water beneath his or her land to the amount necessary for some reasonable and beneficial purpose on the land above the water, such as domestic uses, irrigation, and livestock watering. Wasting water or diverting it off the land was not considered a reasonable beneficial use if such use interfered with the right of adjacent landowners to use the water beneath their own lands for the beneficial use of those lands. While this approach may sound rational, it is not without problems. Most notably, there is often conflict over the definition of "reasonable" use a concept discussed below.

Correlative Rights

Some states, notably California, developed the correlative rights doctrine as an alternative to the absolute ownership doctrine. This approach is an extension of the American rule of reasonable use.

Basically, the correlative rights doctrine recognises the landowner's right to use the water beneath his or her lands but limits that right somewhat by providing that landowners overlying a common source of groundwater have rights to a reasonable amount of that water when the water is applied to a reasonable beneficial use on the land overlying the groundwater basin.

Under the correlative rights scheme, the quantity of a water right is often determined by the amount of acreage owned; there is a correlation between proportions of water supply allocated in times of shortage and the ration of land overlying the aquifer. The problem with this approach, in addition to the "reasonable use" dilemma inherent in the American rule, is that it ignores traditional uses of water and actual need a person holding a smaller piece of land may actually require more water than the law allows, while larger owners may need less. This can lead to hostilities and attempts at gaining water through adverse possession or prescription a practice that often encourage overdrafting.

Prior Appropriation

Most western states have adopted the prior appropriation doctrine. The prior appropriation doctrine simply provides that the first appropriator of water, by putting that water to beneficial use without waste, has a right to continue that use. And such rights are superior to the rights of people who appropriate water at a later date. In prior appropriation states, water rights are usually administered by a state office or official often a state engineer through a permit procedure.

The strength of the doctrine of prior appropriation its permanence and predictability has also become the greatest source of potential conflict. As demographics in the West change and become more urban-based, the doctrine tends to favour those who perfected their water rights in more agrarian times. As we will see below, the degree to which the doctrine of prior appropriation is able to be manipulated and adapted to fir changing times will determine the social, political, and economic future of the West as it relates to water.

Legal Concepts

Besides these four major doctrines that have shaped western water law, two legal concepts also play an important role in the evolution of the law in modern times. The concept of beneficial use repeatedly comes up in the domain of groundwater policy and law. The Utah Code is typical, stating that beneficial use "shall be the basis, the measure and the limit of all rights to use water in this state."

The beneficial use concept was developed during the 19th century to encourage economic efficiency. Although some uses have always been considered beneficial for example, water for domestic purposes or for irrigation, manufacturing, or stock watering outside these traditional uses, conflict and controversy sometimes arise over what constitutes a beneficial use. For example, some state legislatures or courts, such as in Arizona and Colorado, have found water needed for the protection and propagation of fish to be a beneficial use, while others, including Idaho and Utah, have not.

Courts and state legislatures have also been split on the issue of whether or not water necessary for recreation, aesthetic, or scenic uses is a beneficial use of water. This can be a problem in groundwater regulation because of the relationship between surface water and groundwater in streambeds.

Hydraulic continuity, or the interrelationship between surface and groundwater, is a particularly hot topic in the state of Washington, where groundwater permits are increasingly denied because of the impact withdrawals would have on stream flow. If surface water for recreation, aesthetic, or scenic uses is protected, many groundwater permit applications most often of new users of water will have to be denied. While such an approach is environmentally sound, it is not politically popular.

Another issue in the development of groundwater law in some states is whether or not water rights may be acquired by adverse possession. Basically, adverse possession, or adverse use, allows for the acquisition of water rights by the open and notorious use of someone else's water for some statutory period, usually five years. Acquiring rights by adverse use or adverse possession is not unique to water law, but its application in water law in some states has proven problematic.

The application of legal principles in the West may vary significantly even though states may, in theory, follow the same doctrine. New Mexico, for example, follows the prior appropriation doctrine and manages groundwater quite differently from North Dakota, which follows the same doctrine. In New Mexico, where, for all practical purposes, many groundwater basins are virtually nonrecharging, those basins are managed with the understanding that they will eventually be depleted. In contrast, in North Dakota, many groundwater basins do recharge regularly, and that water is managed, for the most part, on a safe-yield basis; that is, the amount of water withdrawn from an aquifer roughly equals the amount of water returning naturally or artificially to the aquifer over an extended period of time.

The Future

To understand groundwater law, it is crucial to consider the social functions water law has served and how changes in the law have mirrored changes in water use and society. Stability of water ownership is essential for economic growth and long-term planning.

Farmers or cities are not likely to build expensive water development facilities if their title to the resource may be called into question in the future. Early in the history of the West, policymakers may have believed that the common law doctrine, or the absolute ownership rule, would provide the stability necessary for long-term planning.

In fact, before competition for water resources intensified, the common law doctrine did provide that stability. However, when competition began to create conflict for groundwater resources, it became clear that one pumper might find the use and enjoyment of his or her groundwater threatened by the activities of pumping on adjacent lands. Converting to the doctrine of prior appropriation, as most western states did, provided the stability necessary for pumpers to understand what their rights were and to plan for the long-term use and development of their water.

This same stability, however, has tended to favour those who acquired their water rights early on, such as irrigators, and to the extent that water laws prevent the transfer and change of ownership of water rights as they do in some states the law has favoured those historical uses and has prevented change in water-use patterns and the development of alternative uses.

So groundwater law has provided the stability necessary for economic growth and expansion. Yet, more recently, that same stability has, in some states, prevented changes in water use and, some would argue, further economic growth and development. Politically, this can lead to pressure for unsound management of the resource. For example, the attempts by some political factions in parts of Washington to deal with groundwater and surface water appropriations separately to accommodate a greater number of uses.

In general, the law has proven itself to be capable of adaptation to change. Some states, however, have lagged behind in responding to changing groundwater use and conditions. The result, in some cases, is that the resource is poorly managed, if managed at all. Most of the states, however, seem to be doing a fairly good job of managing groundwater on a statewide basis. Hawaii, for example, passed a comprehensive water code in 1987, and Arizona's 1980 Groundwater Management Act has led to a substantial decline in statewide overdrafting.

Groundwater law is only one part of groundwater policy. In some states, for example, who gets the water depends more on who can afford to drill the deepest hole or sustain the longest court battle than it does on the formal policy followed in the state. Just as some air polluting corporations can sustain their practice in the face of litigation the benefit of continued operations offsets any penalties or attorney fees some groundwater pumpers may find that the use of the water apart from traditional appropriation systems justifies the cost of a potential court battle.

In other states, in spite of glowing constitutional language about beneficial uses and preventing waste, economic and political realities make it beneficial for farmers and others to waste a great deal of water. Appropriative rights require use of the resource, and water left in the ground for an extended period of time can revert to the state for reallocation to other users.

There is often an incentive to continue using water, even when such withdrawals are unnecessary, to protect an existing right for future use. Many patties are involved in groundwater use and control. Public servants in federal, state, and local water resource and natural resource agencies, legislators and their assistants at all levels of government as well as researchers in a variety of fields, including geology, hydrology, law, public administration, and political science are all among the actors with an interest in groundwater use and management.

Although groundwater has always been an important supply of water in the West and today provides roughly 35 percent of the total water used in the 19 western states, demands on groundwater will increase significantly in the future. Surface water sources in the West are, for the most part, already developed or appropriated. Given federal budget constraints and the resistance of various administrations, beginning with the Carter administration, to building new surface water delivery systems, it is unlikely that major new surface water systems will be built in the future.

Consequently, demands on and interest in groundwater will undoubtedly increase in the future. As demands and interest increase, so will the inherent conflicts that have developed around the distribution of groundwater.

The early and rapid growth of groundwater pumping and agricultural development in western states resulted in the vesting of the majority of water rights to farmers. This laid the groundwork for future conflicts over water use and development, as other users have competed with agriculture for rights to water. Often, the political battles for water have become heated and fierce, and the stakes are high. When Mark Twain commented that "whiskey is for drinking and water is for fighting," he could not have known how fierce the fighting would become. These battles are likely to continue, and the outcome will determine the future of the West.

6

Policy Responses to the Closure of Water Resources

The subject of water allocation and use has gained increasing prominence in the period since 1990, partly because of recognition that some regions, economies and communities have run out of water permanently, or temporarily, at least for some uses. Permanent deficits have become evident in the Middle East region as a whole; temporary shortages have occurred during droughts in the same locations and extreme drought related problems have occurred during the 1990s in southern Africa, in California and even in humid north-west Europe during particularly dry years such as the 1995 summer in the UK. Seasonal shortages are commonplace in semi-arid countries during dry summers, again there are many examples in the Middle East, in for example, Tripoli in Libya, Amman in Jordan or Damascus in Syria. These economies have not been able to supply the modest volumes of water needed to meet the requirements of urban domestic users.

In all discussions about water policy it is important to define the scope and nature of the analysis. Water is a familiar substance and a resource more ancient than the species which depend upon it. However, its role in sustaining life and livelihoods is in many ways much as it was at the dawn of settled agriculture in so far as all humans still need to drink and to eat. And it is the provision of water for crop production which has been and remains the dominant water using activity. With the development of complex industrialised economies, however, the economic significance of water has changed because the shares of food production and the trade in food in domestic and international systems has been progressively diminished.

Agricultural production systems require large volumes of water. Water for crop production comes either from naturally occurring soil moisture or as water delivered from elsewhere in the hydrological system from surface waters or from groundwater. As an economy develops and industrialises, the need for water increases. Industrial activity also requires water, although never in such volumes as does crop production. Industrialised economies progressively reduce the proportional contribution of agriculture to their GDPs; by the late 1990s agriculture contributed only two per cent to the UK GDP; but even in economies which have recently emphasised the importance of agriculture, such as Israel, the figure is only three per cent and for neighbouring Jordan it is only seven per cent. All communities gaining livelihoods in industry and services need large volumes of water to meet their food needs even if their own livelihood uses negligible volumes of water.

The difference between water use in agriculture and in industry is that the returns to water and the livelihoods supported are very different. Industry and services can provide a thousand times more jobs and 20000 times more financial return than would a crop producing enterprise using the same volume of water.

The salience of water issues in a political economy is politically and economically determined. For an advanced industrialised economy water is a small economic element in the overall economy, accounting for a tiny fraction of one per cent of the total economy. Even the huge volumes of water used in agriculture can be available at zero cost for national crop production if the economy is located in humid latitudes.

In poor and subsistence economies the economic options are by contrast very limited. Crop and livestock production are major elements of these economies. The availability of soil moisture is a life and death issue and their are no economic insurances in a subsistence economy, such as would exist in a diverse and widely connected economy with more economically effective water using practices. Meeting the major water needs of subsistence economies is very different from meeting those of industrialised communities in a number of other ways. Sustainable subsistence economies need water to be available locally to meet their immediate drinking water and food production needs. Industrialised economies can access water via trade can also engineer drinking and domestic water supplies from distant sources.

If water is so varied in its role in livelihoods and in its general economic significance it is not surprising that it is difficult to identify policies and policy making procedures which are universally applicable. The purpose of this chapter is to highlight the relationship between the renewable natural resource, water and development. The relationship is a dynamic one and it is essential that it be understood by those evolving and implementing policy. Especially important is it to grasp that the options available to a poor economy are very different to those of a rich and diverse economy. Even more important is it to recognise that it will usually take about 25 years for economic circumstances and above all perceptions and politics to 'develop' to a point where economically and ecologically rational water allocation policies can be implemented.

The discussion will focus on the Middle East as this is the region which has first encountered the problem of running out of water. The region is the first major region in the world, as opposed to a single economy such as Australia or prominent one such as California, to have to consider 'closing' its water resource. It has *not* run out of water with which to develop diverse and sustainable economies; Israel has 'closed' its resource and is at the same time an example of economic vigour. The Middle East has run out of water to service indigenous agricultural sectors with a capacity to meet food needs. It will be shown that policy is evolving according to a predictable relationship between the economic capacity achieved and the water allocation policies that can be implemented. Overall economic development, it will be argued is the most important variable with respect to the development of sound water allocation and management.

The other aspect to which attention will be given is the effect of the global economy on water policy. The political economy of water in the Middle eastern countries is subordinate to the political economy of global trade in food staples. Because so much 'Virtual water' enters the Middle East via trade in cereals the region can remain economically stable. There is a second consequence of the availability of 'virtual water'. 'Virtual water', priced for two decades at much less than its production costs on the world market, has been both a remedy to the Middle east's water deficit, and has at the same time allowed the political leaderships of the region to deemphasise the water predicament. The rhetoric indicates that water problems are being addressed and indigenous water is being used more efficiently. The link between food imports and water is not emphasised and certainly not the unpleasant fact that indigenous water can never meet national food production needs.

It is the water demands for agriculture and plant production, even in the domestic garden, which are the uses which cause economic crises and social stress. The reason is that the needs of an individual for food are relatively inelastic. We each need about 1000 tonnes of water (cubic metres) per year to raise our food needs. We only need about one cubic metre of water to drink per year. The figure for domestic use per head can vary from as little as 3.5 cubic metres per year (10 litres per day) to a very high figure indeed-even 2000 cubic metres per year-if a family has a large garden in an arid or semi-arid country where people enjoy a high standard of living. At the same time the increased standard of living of industrialised communities bring a propensity to use more water for domestic purposes.

In any discussion of water management policy it is important to emphasise that water management has a number of dimensions which tend to be de-emphasised. Water management is concerned with improving access to, and the efficient use of, water. Such measures will increase the volume of goods produced by increasing the availability of water and will also improve its productivity by increasing the returns to water by, for example, applying water more efficiently to irrigated crops. Just as important as these measures, however, is the effective allocation of water. This book bears the title-allocation and management-in order to emphasise the prime importance of allocation in any water management policy. Water use efficiency can just as surely be achieved by allocative measures as by investing in technology to improve productive efficiency. The problem with any allocation policy is that it requires that political as well as economic priorities be taken into account. Reallocation normally means that one party will lose the resource in order hat it be made available to another user. That the economy as a whole might benefit from reallocation does not impress existing users with millennia of tradition and status invested in their way of life.

Phases in Water Allocation and Management in a Dry Region

The long experience of managing a water surplus in the Middle East up to 1970 was an important and in some ways unfortunate conditioning experience for water managers in the region. Scientists, officials and politicians developed an ethos that new water would always provide a solution provided that enough engineering ingenuity could be mustered. By the 1970s supply management ceased to be an option for much of the region.. However, because of the availability of international 'Virtual water' it has been possible for the Middle Eastern and North African economies to balance their water budgets. The last quarter of the twentieth century has been a transitional period during which the region's water managers have been able to meet water needs without addressing the management of indigenous water with radical demand management practices. Demand management includes measures such as reducing water leakages in distribution systems, 'closing' the water resource for activities and sectors which bring a poor economic return to water and re-using urban and industrial waste water. Most Middle Eastern economies should have begun to consider 'closing' their water resource by the 1970s.

Chatterton emphasises that a state authority must take a number of steps with respect to the ownership of water, including the identification of water rights. They suggest that it is very important to identify the water rights that would exist at 'closure' and the different rights that would exist for any subsequent allocation. Most important they claim is the identification of a budgeted overall allocation which must be less than existing and new water rights so that any reasonable appeals could be met. That attitudes to the value of water would change significantly with a water closure policy is inevitable and useful. That the politics of the announcement and implementation of a water closure policy would be intense goes without saying as does the need for robust institutions through which the interests of diverse and antagonistic users can be mediated.

Figure: *Strategy for managing water resources in the dry American West.*

The only economy in the Middle East which has been in the position of economic strength and institutional robustness to take the political step of announcing a need to 'close' its water resource is Israel. The realisation that closure was necessary did not take root quickly. The notion was identified in 1963 by US officials and by Israeli scientists and officials. But it required three sets of circumstances to come together in 1986, almost a quarter of a century later, for the policy of water closure to be politically feasible. It had been economically feasible for over a decade but economic feasibility is almost always much less important in a political economy than political expediency. The three circumstances were:

- a ten or more year discourse prior to 1986 on the issue of water use efficiency to which Israeli scientists and later Government officials contributed often vigorously.
- pressure from the US Government that Israel should get its economy into some better order with fewer subsidies. Water policy was just one aspect to be addressed. The sanction for not taking measures would be that a US$ 10 billion grant would be with held.
- a timely drought in 1986 made it possible to communicate the seriousness of the water crisis to the Israeli public including the heavy water using interests in agriculture-using over 70 per cent of national water but only three per cent of the GDP and the water effective sectors such as industry and services-using only five per cent of national water but contributing 97 per cent of the GDP.

That 1986 was a moment in history of Israel's political economy at which it was possible to impose a water closure policy fits with some other general ideas in currency in environmental and development theory. Karshenas has demonstrated the relationship between environmental resource using practices at early and later phases of economic and social development. The tendency at an early phase of development is to use and degrade renewable natural resources such as water. At a later stage a more diverse and strong political economy can contemplate the reconstruction of resources negatively impacted by earlier phases of development.

Grossman and Krueger have suggested that a similar concept of the environmental 'Kuznets curve' (EK-curve) is a useful one. They draw attention to the similarity of the favourable trend in resource managing policies, of for example pollution, which appear to be associated with the achievement of a diverse and strong economy. In addition to arguing that economic growth is good for the environment they also suggest, more controversially, that the liberalisation of international trade can reduce aggregate environmental degradation. Ferguson et al find that the EK-curve 'can say little about whether a country's economic trajectory yields a greater or reduced aggregate environmental effect' and they agree with the World Bank that 'the existence of an environmental Kuznets curve, and this point cannot be emphasised enough, does not imply that countries will naturally grow their way out of environmental problems'.

The Ferguson report insists that the shape of the EK-curve is 'policy determined' and that 'the nature of the relationship between economic growth and the environment lies largely in the hands of those responsible for environmental policy and its enforcement.' They also insist that sound futures can only be achieved through a 'timely and closer integration of economic and environmental policy-making'.

The Israeli case appears to fit both models as well as the notions of their critics. Its economy achieved rapid industrialisation in about 25 years between the mid-1950s and the early 1980s which would according to those finding explanation in the environmental Kuznets curve provide the conditions which enabled Israel's policy makers to take what appears to be a sound economic and environmental position on water allocation by 1986. The Israeli experience could also be taken to fit Ferguson's argument that what is required for sound policy is the existence of a political will to implement environmental policy inspired by an integration of economic and environmental principles. However, to insist that a principled environmental policy, in this case on water, could have come early in the process of industrialisation rather than late appears to fly in the face of the evidence of Israeli development history. It also contradicts basic notions on how policy determining scientific and political discourses evolve. It is evident that approaches inspired by sound economic, demand management, principles and sustainable environmental principles came after supply management approaches were no longer tenable. That these sound principles have only been thoroughly adopted in the economy in the Middle East which has successfully and thoroughly diversified, namely Israel, is significant. In the other political economies the scientific discourses and hydropolitical discourses are in train.

International Context of Domestic Water Policy

The main purpose of the chapter is to demonstrate that it is essential that those who attempt to provide an economic perspective as an input to policy making recognise that water management takes place in open political economies and not in closed hydrological and engineered systems. Policy makers in the real world seek effective operational systems with which to solve their problems and will prefer the remedies to be found in such systems rather than to seek solutions in intractable legal wrangling, in unrewarding negotiations with unyielding neighbouring riparians or in politically stressful water re-allocation at home. Policy makers in water short economies seek remedies to their water deficits in the wider global economy rather than the hydrologically defined one which constrains them. Trade provides the remedy.

Figure: *New water policy*

It has been shown that the option of moving highly subsidised 'Virtual water', that is the water contained for example in cereal imports, has proved to be an extremely effective and economically efficient method of meeting water deficits. Those responsible for developing strategic policy on water will have to look closely at the availability of global water, at the performance of key economies in the global trade in cereals, both suppliers and potential consumers, and at the capacity of economies currently self-sufficient in water for agricultural production to remain self-sufficient in food. They will also have to understand thoroughly the emerging international institutions which will affect trade in 'Virtual water'. The political economy of water in a particular country or region will remain subordinate to the political economy of global trade in cereals.

The availability of 'Virtual water' is of major significance in the domestic water policies of water short economies. It allows such economies to balance their water budgets without the political stress of confronting intransigent water using interests at home with the need to improve their water use efficiency. 'Virtual water' provides the ideal 'invisible solution' and allows policy makers and officials to defer painful measures such as closing the water resource.

They can sustain the rhetoric of water and food self-sufficiency which accords well with the widely held traditional views on the importance of the agricultural sector in the economy as well as with the interests of the usually large rural population. Policy makers can continue to defer addressing the national water predicament.

They have done so in the Middle East for the past 25 years in a transitional phase in which not only has 'virtual water' been available, but priced advantageously through subsidies born by producers in industrialised countries. Wheat was traded at half its production cost as a result of the USDA and the EC competing down of world grain prices.

The founding of the World Trade Organisation in the spring of 1995, along with a tightening of international grain reserves and evidence that China would be on the market in 1995 and subsequently, had a dramatic impact on world grain prices which doubled in the following 12 months.

Despite this change in the prices 'Virtual water' still remains a very good deal within such transactions. There is still no engineering measures which could mobilise the between 20 and 30 billion cubic metres per year of water needed to produce the grain being imported annually into the Middle East. Demographic circumstances dictate that the Middle East's food and water gaps can only increase in future. Meanwhile current and foreseeable technologies cannot provide water at costs which can be accommodated by crop production systems.

Evolving International Economic and Institutional Contexts and Water Policy

New circumstances which face importers of 'virtual water' as the result of the shifts in policy on agricultural production in the US and the EU following the establishment of the World Trade Organisation in 1995. The uncomfortable change in trading circumstances during 1995 has already sharpened awareness of the need to improve water use efficiency and make the best possible use of indigenous water.

The Governments and peoples of the water deficient Middle East have to face the unpleasant challenge of increased staple grain prices at a point when their economies are either weakened by the price of oil. During the 1990s crude oil has only been able to command a price lower in real terms than its 1972 level. The nonoil economies face even more serious balance of payments problems. The interpretation of whether the resource poor economies of the semi-arid Middle East can meet the additional burden of higher world grain prices depends on whether the observer tends to see the position s being 'half-full' or 'half-empty'. There is considerable evidence that coping with the ceaseless progressive demand of additional population is as much as these economies can expect to achieve. On the other hand the structural adjustments being forced on economies such as Egypt are bringing about change and inward investment is taking place.

A discussion of the future development prospects of a regional economy may seem to be distant from the making of water policy. It cannot be emphasised enough that the diversity and strength of an economy in the semi-arid Middle East is the essential feature to be considered. The strength of the economy and its capacity to trade in the world market and especially in the international grain market is a basic element in national economic and strategic security in all the countries of the region especially with respect to water. Access to 'Virtual water' will also make it possible to pace the introduction of water closure policies to accord with the rate at which the associated discourses and politics can be steered. The recent increase in grain prices could act as a catalyst to encourage the grain importers of the Middle East and North Africa, who together comprise a bloc of pivotal importers, to act together with a coherent policy to create favourable trading circumstances. Becoming expert in grain futures and even investing in secure supplies of grain from regions which enjoy a comparative advantage in soil moisture availability should be a widely recognised priority,

Water policy is the concern of this publication. The aim of *this* contribution has been to show that there are a number of factors and pressures on policy makers at the macro, national, level which determine whether sound ideas can or need to be heard by the Government policy making community. The good ideas may be economically or environmentally very highly principled, but when put into scientific, institutional, political and media discourses they will be subordinate to the political imperatives preoccupying those in Government. Policy makers will be concerned with retaining power and with encouraging the circulation of ideas and proposals which provide opposition groups, official or unofficial, with information which could endanger their political position. That scientists and consultants recommend policies that are environmentally sound, technically effective and economically efficient according to the internationally acclaimed principles, for example of intergenerational transfers, is of little importance if the proposals contradict immediate political imperatives. Outsiders drawing attention to weaknesses in the national economic fabric, such as the inadequacy of water supplies, does not accord with Government sanctioned discourses and attempts to focus attention on such issues will be unwelcome. Notions of food and water insecurity are not therefore heard or encouraged, and if they do enter the discourse they are opposed with rich rhetoric or more probably ignored.

Having established the difficulty of having an impact on influential discourses it has also been shown that the political economy of water and related water policy are subject to economic factors. There has been an attempt in the paper to show that the Karshenas model of environment and development, and the environmental Kuznets curve, do help in understanding the inevitably lengthy process which shapes water policy. A state authority can only implement politically stressful measures such a 'closing' a water resource when an economy is diverse and strong. A diverse and strong economy takes a minimum of 25 years to evolve. Governments running economies which have not achieved this status will go to great lengths to avoid taking measures which are economically and environmentally rational. It is politically easy in such circumstances to argue that the sound economic and environmental polices are unaffordable.

Management Approach to Water Scarcity

Water scarcity is the result of pressures from population growth, of environmental change and degradation and unequal distribution of water resources. With increasing competition for limited supplies, conflicts among water users intensify. Water scarcity and resultant conflict are often consequences of inappropriate or inadequate management frameworks, linked to general economic policy and Government structures, with solutions depending on national and regional geo-political realities. Shared, transboundary water resources and import of water are realities in water scarcity management. For example, an overview of water resources in SADC countries highlighted the regional ramifications of water scarcity, and indicated the many national and supranational issues and potential conflicts associated with sharing resources and major transfers between river basins and countries.

Figure: *As the city experiences water scarcity in some parts and launches*

Economic theory predicts that when water becomes scarce it would be expected that water prices would increase to reflect demand pressures, and that water users would save and restrict consumption and other use to essential social and high-value uses. The market would see to it that the marginal benefit unit is the same for all uses and that therefore there is no gain from re-allocation. Prices would then cover the cost of supply to ensure continued availability. However water pricing and management and water markets are hampered by legal and institutional constraints, and traditional and regional attitudes to water, together with high transaction costs and physical limitations in handling the commodity. A system of firm and transferable rights to specified quantities of water that can be exchanged among alternative users through decentralised market processes could, however, work in a well developed economy, while it is argued developing countries require more Government control under assorted water permit systems.

The extent to which a country is vulnerable to water depends not just on the available water, its distribution over time and its quality, but more on the extent that the supplies can balance and sustain present and future demands. Climate is the principal factor in water scarcity, but population and economic development are the main influences on quality and demand that together define scarcity as the level of competition between uses. Water starts to become vulnerable when annual internal renewable water resources are less than a critical value, often being cited as 1000 m per capita. Specific indicators have been proposed to reflect factors such as regularity of supplies, external dependence, demand pressures, non-sustainable water use, and level of wastewater discharge, as well as water utilisation for drinking and irrigation. However, contrary to water vulnerability, scarcity occurs both in dry and humid climatic conditions, influenced by many interlinked factors, including growing water demands from economic development and inadequate management due to institutional shortcomings. While scarcity and competition for use of water are basic factors, the main issue is the institutional and economic capacity of society to manage water scarcity and make water available and suitable, in time and space, for different uses and

The Harvard Middle East Water Project (HMEWP)

HMEWP proposed an economic approach to optimal social allocation of water. Water values were estimated not only as the prices in the free market but also reflecting national social aims involved in water. The cost of supplying and transporting water was estimated, and the private demand curves for households, industry and agriculture were established. Finally, national water policy was expressed as additionai demand for water at different prices, and the equilibrium prices that balance supply and demand were calculated. The economic Approach transferred the contentious issue of sharing of scarce water into a negotiation of monetary values secure efficient use and harmonious relations between different users.

From this aspect, the approach to water scarcity management is closely linked to the capacities of the national economy and institutions. While developed economies have the necessary economic potential to plan for and adapt to water scarcity, developing economies, generally subject to and constrained by high population growth, might not have the capacity and economic diversity to manage water scarcity. The main issue in water scarcity is time and the differences in management of water scarcity in developed and developing economies. Studies of major drought events in the 1970s and early 1980s showed significant decline in GNP and domestic savings, worsening balances of payments, and loss of job opportunities. Water scarcity is, however, an irreversible process, affecting national economies over long time spans, and there is need to alert decision-makers to the related threats to the economy and the state. There is therefore a need to assess the social and economic costs to society of constrained and delayed social and economic development due to water scarcity.

With these dimensions, water scarcity becomes a threat that could contribute to general failure of weak developing country economies, and water scarcity has become an issue of national security. Decision-makers must be made aware of the social and economic impacts of water scarcity and the importance for the stability of the state and the economy, in order to mobilise the necessary political support for development of effective and timely measures.

Water and National Security

Water scarcity is increasingly seen as a cause of domestic and international conflict, and therefore a risk to national stability and security. This preoccupation is reflected in an increasing number of water-scarce countries, where water is being included in the agendas of the highest authorities responsible for national security. Similarly, water scarcity and its importance for national security are being given increasing priority by international donors. Water scarcity as a national security issue is developing into a useful concept for approaching the problem and in attracting attention at the political level to the importance of water scarcity as a long-term, macro-economic and sustainability issue. The security risks, deriving from environmental scarcity, population growth and unequal distribution of water, appear as reduced state capacity, and also civil violence, rather than international conflict. Water scarcity could contribute to frustration of weak national administrations, leading to breakdown of the state as water scarcity becomes a major threat in developing countries trying to achieve their social development objectives. As a consequence, when perceived as a threat to social and societal security, water scarcity should mobilise the attention of national policy-makers and the international community sufficiently soon to address the problems before they reach uncontrollable dimensions.

Water and Food Security

As a result of increasing demands, the global food gap is widening. The price of food in international markets is projected to increase significantly, and weak economies are already facing increasing difficulties in buying food. At present, 2400 million people depend on irrigated agriculture for jobs, food and income. Over the next 30 years, an estimated 80% of the additional food supplies required to feed the world will depend on irrigation. Irrigated agriculture will be expected to produce much more food in the future, while using less water than it uses today.

Environment and Health

Environmental concerns are increasingly shaping water management plans. The environmental impacts of schemes to supply and use water can have a massive hydrological impact, and impose water scarcity on other users and future generations. Crucial factors, such as minimum flows in rivers, will gain in importance and it is likely that the value that society places on water for environmental and health purposes will increase in future.

Management of National Water Scarcity

Recognition of water scarcity at an early stage in planning forms the underlying principle for any action towards water sector policy reform. Action is, however, often reactive and short term, and policy reforms and water acts are more easily accepted by the public and at the political level in crisis periods of drought and acute water shortage. Many countries, especially in humid zones, maintain a position of no problem and so no need for management, and hence react slowly, often not until water shortage or water quality degradation is constraining national social and economic development. Approaches to water scarcity management need to be based on political recognition and general awareness of the realities of scarcity as a major national problem. While there is reluctance to deal with issues that risk being contentious, and so results in deferral of Water policy in Egypt and Mexico.

Egypt: National strategies include water storage in coastal lakes, desalination, water conservation in the upper Nile, re-use of irrigation drainage water, increased irrig utilisation, and modified cropping patterns

Mexico: An alternative to eco Government control under waternew 1992 Federal Water Act and developing a unified water abstraction and wastewater disposal perm experimental basis in an area demands appropriate measures, the risks are accumulating of catastrophic social and economic consequences, including deteriorating food security with resultant conflict and possible collapse of state and customary institutions, thus threatening national security. Ultimately, it will be the countries themselves that will have to find their own means of adjusting to changing circumstances, or they will collapse.

Of crucial importance is that many policy measures-both traditional and of newer provenance-for operational water resources management, short-term planning and conflict resolution to free and provide more water are narrow, and, even under ideal institutional setups, will not work, and can even in time provide misdirections in addressing national water scarcity. The requirement is for a broad, long-term approach, based on national security and economic growth objectives, well consolidated with national development, economic reform and macro-policy, as part of a larger strategy of long-term societal change to counteract the root causes of effects tending toward breakdown of the state.

Misconceptions on Solutions to Water Crisis

Frederiksen, in a review of common misconceptions concerning the emerging water crisis, soliciting more real-world and down to business approaches, points out basic constraints and misconceptions, and caveats concerning current management approaches.

Scarcity of Time to Act to Meet the Needs

There is a necessity to act, even on controversial issues, and therefore to use established measures rather than to wait for new approaches that might not be implementable with available resources and capacities.

The Macro-economy and Water Scarcity in Yemen

Yemen is facing economic stagflation and difficulties with budget and balance of payments. Current development practice is to deplete natural resources assets, notably water, to finance consumption, unproductive industrial capital and poor savings. Low returns on investments and external indebtedness shift resources towards current use, reducing water sustainability. The losses of water assets are not recorded under current accounting practices. With capital resources injected into the import sector and inefficient production, this is not an adequate source of derive from water-intensive economic growth which results in resource deterioration, with economic interactions between mobilisation of water resources and the macro-economic and sectoral policies. The balance between macroeconomic and sectorial development and sustainable protection of water resources needs to be determined as structural and policy reforms are implemented and the economy evolves.

Common Misconceptions on How to Free and Produce Needed Water

Water availability is often expressed in terms of total water resources, while only the utilisable portion of the water resources can meet the needs of a nation, and there are few options for increasing this portion, such as regulation of surface flows and conjunctive use of surface and groundwater.

Diverting a share from agriculture to higher-value uses is often quoted as the solution to water scarcity. However the scope for re-allocation is often limited to the low-flow season, and does not include high-flow season water, when major cropping often takes place in irrigated areas. Water acquisitions from agriculture for municipal and industrial uses tend to be localised, and thereby reduce the economic basis in the immediate vicinities of cities, at a social cost in lost production that might be larger than the losses from reservoir inundation.

Improved agricultural efficiency is often stated as an obvious option to save substantial quantities of water for other uses. However, irrigation projects pass unused losses to downstream users, and, as a result, basin efficiencies are generally high. Thus improved agricultural efficiency could lead to reduced supplies and potential water conflict among downstream users, and conflict with national and social and economic interests as current production might be reduced and exchanged for new production, at a cost but with no net gain. Water should, however, be saved in urban coastal cities which have low water use efficiencies of 10-15% and where return flows are lost into the sea, and in irrigation areas that discharge return flows into saline depressions. In national water scarcity management it is the basin efficiency that is relevant. Demand management and water pricing are not likely to free large quantities of agricultural water to other uses since water is already managed as a main constraint to cropping in the water-short, low-flow season. In addition to costly institutions and physical infrastructure needed for water markets and trading, these policies will only re-allocate water that is already used in a productive way, and is likely be counterproductive to poverty alleviation and national social and security objectives. Transfers and acquisition of water rights need to be closely linked to land rights.

As pollution control, together with recycling and desalination, cannot offer inexpensive and rapid options, and development of groundwater and natural flows arc limited, one of the few option is to develop surface water storage. Improved cropping management, including production-effective and high-tolerance crops, and the use of water of marginal quality, also have a lot of promise.

Policies to Mitigate Shortage of Funding

The major concern is that while only short-term, and often insufficient, funding is being allocated and made available to the water sector, the very large financial needs in the next decades have not even been analysed and prioritised. Privatisation, which is often being promoted, remains dependent on the traditional source of funds-Government or beneficiaries-and has proven to be an uncertain source of financing, especially under conditions of water scarcity and economic recession. In contrast, increased service fees and transfers of financial responsibilities to beneficiaries represents an efficient and timely option for improving funding, with major longer-term retrenchment of financial responsibilities from Government. It is consistent with, and goes in parallel with, the slower process of removal of subsidies in the water sector.

With changing social and economic conditions and weak governance in developing countries, what are considered normal conditions for water resources and the capacity to manage water scarcity can be expected to change drastically in the next decades. Current conceptions and approaches to managing water scarcity focus on use of water where it is available, and re-allocation of available water for its greatest benefits. While detailed data on water scarcity and its impacts are generally absent, there is evidence that these conceptions and the focus on demand management and non-structural policy instruments are insufficient, misguiding and delaying to more effective action because they do not reflect the requirement to manage water scarcity as part of the larger process of societal change. This is especially true in developing countries with weak economies and unstable state formation. Traditional approaches to water management will therefore not be sufficient in situations of extreme national water scarcity, possibly resulting in the collapse of state and customary institutions. Hence water scarcity management needs to be given a broader scope and made part of national strategies for societal and economic reform. To address these challenges, FAO is currently developing a guide to provide recommendations to member countries facing current or potential water scarcity.

The Water Context

Historically, water has been a grossly undervalued and often neglected resource. Slowly this is beginning to change as water scarcity, deteriorating water quality and rising marginal costs of supply have brought discussion of water management into the political arena and begun to shift perceptions of water from being an entitlement to an economic commodity. As economic scarcity and its recognition increases, so does the threat of conflict over the allocation and distribution of water at local, national and international levels.

The involvement of the state in the water sector in many countries has also recently been subject to criticism. Firstly, the range and fragmentation of Governmental policy interests vested in the sector have acted against coherent and consistent policy development and public sector investment (e.g. water quantity and quality issues, and surface and groundwater management, have been insufficiently integrated). Secondly, considerations of economic efficiency have been marginalised, and states have followed a supply management orientation, often in the post war era with emphasis on large engineering structures. Until recently, relatively little attention has been paid to improving water use efficiency, the use of pricing and economic incentives, or to encouragement of small-scale measures for local management and supply augmentation.

The Groundwater Problem

Groundwater, even more than surface water, has suffered from undervaluation and neglect. This, one suspects, is a function of its invisibility ('out of sight out of mind'), lack of knowledge and understanding, inherent difficulty of study (or model), and the small scale of much exploitation. Under both state and private development there is evidence of nonsustainable management and groundwater resource degradation in different parts of the world. However depletion is still insufficiently recognised as a major global problem and seldom is its human significance assessed. Similarly, groundwater quality issues until recently have taken a back seat in most debates about environmental degradation. The consequences of such neglect are serious: groundwater, though naturally suited to human consumption, is vulnerable to contamination and hard and expensive to clean up.

Aquifer Functions

The functions of aquifers in supporting human productive and livelihood activities are considerable and diverse. As an essential part of the hydrological cycle, aquifers collect and store surface runoff that might otherwise be lost to the sea. Thus they provide natural storage of fresh water in a form that can be readily utilised by society when required. Where recharge is adequate the resource can be managed sustainably, and be used as a buffer in periods of deficit or drought, or retained for use by future generations. Other beneficial functions of aquifers include, depending on location and geology, their role in assimilation of waste, in sustaining wetlands of ecological interest, and maintaining springs and surface water supplies; they may also provide a natural storage reservoir for water surplus to immediate requirements, and be used for artificial recharge. Aquifer exploitation may require relatively little investment and is potentially available to any resident with overlying land, though also depends on prevailing institutional factors. Technological development has recently extended the volume of groundwater within economic range, but at the same time has increased its susceptibility to depletion.

Groundwater Development

Because of its physical characteristics, groundwater is a common pool resource typically developed on a small scale, often privately by individuals and communities. Where open access ('no property') situations exist, with exploitation open to multiple users, individual rationality has often led to unsustainable competition not beneficial to society as a whole. Beyond a point, groundwater resources (GWR) are subtractible, and where there is no institutional control over sinking wells, nonsustainable exploitation can take place-the so called 'tragedy of the commons' scenario.

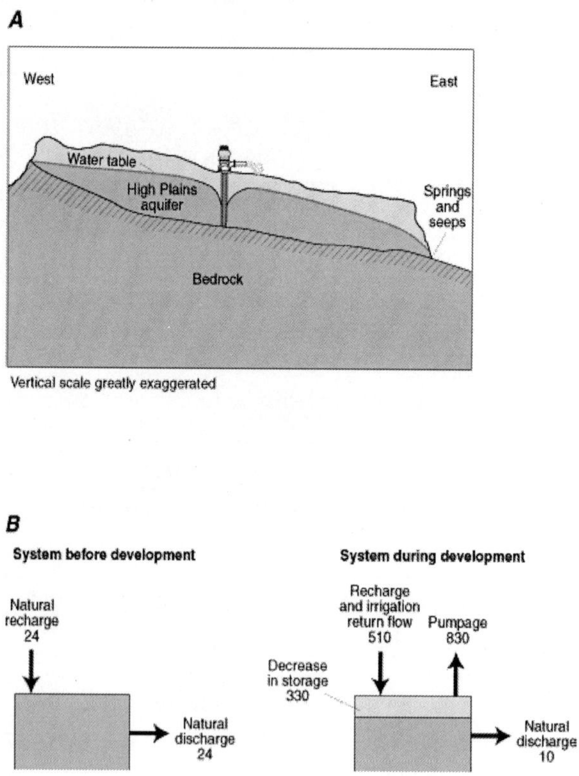

Figure: *The effects of ground-water withdrawals on the southern High Plains aquifer*

Not all GWR resources, however, have been developed in an uncontrolled or laissez-faire manner. In this chapter we have selected two case studies for examination that portray contrasting institutional situations: the Gaza Strip in Palestine, where management systems have failed to keep pace with socio-economic and political developments and a virtual 'pumping race' has emerged; and Yao Ba, an irrigation scheme in Inner Mongolia with a tradition of tight centralised regulation and control. As we will see, however, certain GWR degradation problems have arisen in both locations and circumstances: even where abstraction is controlled by the state, nonsustainable exploitation can occur.

Groundwater Degradation

A general definition of GWR degradation might be 'any deterioration in the stock or quality of groundwater resources or systems'. It can occur for natural or anthropogenic reasons, although we are here primarily concerned with degradation induced by man. It can be permanent or reversible, though the cost of reversal (or treatment) may be high. In all cases, however, it has implications for the availability and cost of freshwater supplies, and for meeting the needs of future generations.

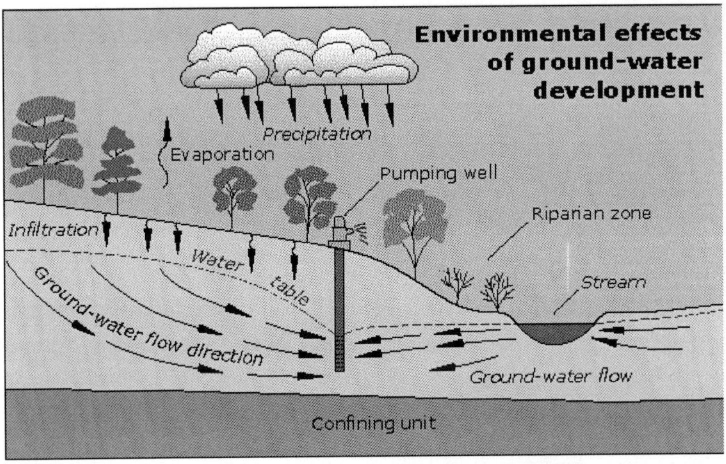

Figure: *Environmental effects of ground-water development.*

A convenient framework in which to consider the symptoms of degradation is to classify them in three categories: depletion; depletion-induced; and contamination. *Depletion* relates to a reduction in the quantity of the resource as indicated by a fall in water levels. *Depletion-induced* relates to secondary changes induced by the fall in water levels, particularly saline intrusion, land subsidence, and reduction of water storage capacity. *Contamination* relates to the loading of contaminants (pollutants) at a rate exceeding the aquifer's ability to assimilate, and is independent of depletion. A fourth degradation category, augmentation or raising of water levels, could be added as this may result in soil salinisation and water logging but this aspect is not pursued here.

Contamination can similarly be broken down into categories, referred to as point source and diffuse (or nonpoint) pollution. Point source pollution is traceable to a particular site or entry point, and often emanates from industrial effluent or municipal waste. Diffuse pollution is that which cannot be traced to specific sites or locations, and is typically associated with intensive agriculture, particularly the use of nitrogenous fertilisers, but also may be atmospheric in origin (e.g. acid rain). Pollutants include hydrocarbons, heavy metals, organic and bacterial compounds, nitrates and pesticides. Their presence in GWR may reduce useable water stocks, downgrade their value (e.g. change from drinking to industrial water quality), introduce health risks, and increase supply cost. In sum, degradation can be considered as having both quantitative and qualitative aspects. This is most conveniently handled using the economic concept of value. In economic terms, degradation represents a reduction in real value of a resource stock as a consequence of deterioration in either its quantity or quality, or indeed the two together. Unless otherwise specified, we henceforth use the term degradation in this general way.

A Conceptual Framework

From a broader social science perspective, GWR degradation is part of a circular cause and effect process. Degradation is viewed as the result of the operation of economic and demographic processes that over time serve to increase the demand for groundwater within the economy and the level of pollutant (or load) reaching the aquifer system. These processes are communicated to the resource via a matrix of institutional factors which serve either to limit or intensify the impact on the groundwater resource; they are defined generally to include any factor that may influence pressures on GWR, and may be legal, cultural, organisational, technological or relate to the political system. The severity of degradation is in turn influenced by the specific aquifer characteristics (such as volume, depth, heterogeneity) that limit or raise the vulnerability of the resource to degradation. Thus both institutional and physical factors govern the susceptibility and nature of GWR degradation in any particular situation.

Once GWR have begun to degrade, effects will start to show in the human productive and livelihood systems they support. These impacts may be economic, environmental or social in nature-including, for example, increased operating costs, changes in cropping pattern, infrastructural damage, and damage to health-and be differentially distributed in society. The nature and extent of the socio-economic impact of GWR degradation in a given situation are subject to a similar set of location-specific institutional factors. Analysis to establish these linkages requires interdisciplinary effort that examines biophysical and human variables in a 'web' of cause-and-effect relationships. Although the process is cyclical, with no start or end point, the cycle is never constant and feedback and mutation enter the system. In this way factors can be both cause and effect at the same time.

Mitigation and Management

Most theoretical examination of optimal natural resource exploitation policy treat resources as either *exhaustible* or *renewable* but in practice the line between the two may not be easily drawn. Groundwater is an important example of a resource that may or may not be renewable depending on physical circumstance and strategic management. The resource can be managed in one of three broad ways:

- degrade the aquifer at present or accelerated rates
- maintain the aquifer by reducing abstraction and contamination to sustainable levels
- restore the resource to original levels.

The choice of strategic option relates to basic questions of time-preference and the future value of the resource stock. Methodologies for selecting options have been discussed at a theoretical level but not to our knowledge in relation to groundwater. Good policy demands making difficult temporal comparisons requiring detailed local knowledge and analysis but cannot preclude judgement. In practice, selection commonly occurs not as a result of analysis of likely costs and benefits to aggregate society, and to the various stakeholders within it, but often by default.

In practice the strategic options are not as stark as portrayed above and there are other important aspects of strategic management. A convenient frame of analysis is to group them into three categories, supply-side measures, demand-side measures and intermediate measures. *Supply-side* measures relate to the augmentation of existing water sources and normally involve investing in physical structures to meet a projected water 'demand curve' or to rectify a pollution problem.

This has been the traditional and engineering-led approach to strategic planning of the water sector. *Demand-side* measures, on the other hand, are designed to reduce the use of water, and improve use-efficiency, through economic incentive mechanisms or by regulation. Incentives can be direct, through the use of volumetric pricing (paying for quantities of water used), or indirect through taxation or pricing of agricultural inputs.

Intermediate measures are designed to save water by reducing wastage, such as the introduction of more water-efficient irrigation technologies and the lining of distribution canals, though only high value usage may be able to cover implementation costs. Where sufficient power is invested in the state, it may also be possible to divert water from low value usage, often thought of (not necessarily correctly) as irrigated agriculture, to higher value industrial or domestic consumption.

Two Case Studies

As part of our current ESRC research programme, we have conducted nine case studies of groundwater degradation problems in China, India, Thailand, Mexico, Ghana, Jersey and the Palestinian territories, representing a range of political, socioeconomic and hydrogeological circumstances. Here we turn to site specific discussion of degradation problems and management options in two contrasting study areas.

Yao Ba Oasis

Yao Ba is situated in an arid pastoral area bordered by mountains to the east and the Gobi desert to the west, in the autonomous region of Inner Mongolia. About 35 km in size, its location is determined by an aquifer which underlies the area at a depth varying between 15 and 60 metres. An irrigation scheme was established in the area in 1972, since when it has become an oasis of prosperity and home to a population of over 7,000 people. Today its thriving economy is based on the irrigated production of wheat and other cereals for human consumption and livestock feed. Farmers are composed of local villagers who have turned from pastoralism to sedentary fanning and settlers from impoverished and overpopulated provinces who have migrated to the area. The scheme now has 250 functioning boreholes powered by electricity which gravity-feed numerous individually operated irrigation basins.

The Problem

In the last six or seven years, prosperity has been threatened by the problem of rising salinity in the desert margin, a problem that will grow at present or even reduced abstraction. Changes in hydraulic gradients resulting from aquifer drawdown have led to the intrusion of saline water into the aquifer from a nearby (ephemeral) alkaline lake. Yao Ba thus demonstrates a *depletion-induced* problem which has begun to salinise soils, reduce crop yields, remove salt-sensitive crops, and in some cases force the abandonment of wells. The problem facing the authorities is how best to handle this physical problem and minimise its socio-economic impact on the local community.

The Management System

As in the rest of China, water resources are the property of the state. In Yao Ba, water policy emanates from multiple levels of Government including the central Government in Beijing, the Government of Inner Mongolia, the Alashan League division, and the local banner or county. Day to day management of the irrigation scheme is shared between various bureaux of the banner, together with the Yao Ba administration, local administrative units called gatchas, state farms, and farmer groups. The Electricity Authority also plays an indirect if unconscious role in water management through its charging mechanisms.

The establishment of boreholes in the area is under the strict control of the water authorities, partly through licensing and partly through supplying drilling machinery and credit. While resource ownership is still firmly in the hands of the state, scheme operation is in the process of decentralisation (the 'household responsibility system').

In practical terms boreholes are now the property of the gatchas, the most local level of Government, and their management rests with farmer user-groups. Utilisation of the water itself is the responsibility of individual farmers who, since 1994, have been able to make their own cropping and water allocation decisions. Water is distributed to farmers on a strict rota basis and is virtually unpriced. However electricity charges for pumping have in the last two years been raised several-fold and their impact is beginning to affect farmers' water-allocation decisions. Decisions relating to further development or abandonment of the area's groundwater resources, however, rest largely with the Water Utilisation and Conservancy Bureau. The Bureau has responded to the growing salinity problem by restricting borehole establishment which has effectively curtailed expansion of the scheme.

Strategic Options

Conceptually, as noted above, the strategic options are to *degrade, maintain* or *restore.* At the existing rate of pumping (about 30 mm/yr), drawdown and thereby salinity will increase over time and ultimately destroy the aquifer. While having immediate advantages, in the longer term this would have an enormous economic and social impact on local livelihood systems, and more widely on the supply of cereals and livestock feed in Inner Mongolia. On the assumption that annual recharge is of the order of only 6 mm/yr, the introduction of a fully sustainable policy will (*ceteris paribus*) would lead to a drastic reduction in irrigated area and have a massive impact on livelihoods. Knowledge of aquifer recharge levels is very imperfect and for this reason there are serious practical problems associated with sustainable management. The third strategic option, aquifer restoration, merits little consideration here for social and economic reasons, demanding as it would abandonment of the scheme and the loss of livelihoods by many. In practice, therefore, the realistic options involve a balance between the first two of these strategies.

Fortunately the position can be improved using the supply, demand and intermediate framework identified above. In Yao Ba there are few obvious supplyside measures although the drilling of interceptor wells to reduce saline accumulation has been proposed. On the other hand, there appears to be a considerable potential for reducing water demand through economic or pricing mechanisms, though implementation would not be straightforward; proposals introduced by the Bureau in 1994 were not implemented following a national decree from Beijing intended to protect fanners.

Fortunately there appears to be ample potential for improving water use efficiency at both scheme and farmer levels. These include the lining of channels, replacement of basin irrigation by micro-irrigation systems, the use of plastic membranes to reduce evaporation, and changes in cropping pattern. It would also be theoretically possible to abandon the saline area and move farmers elsewhere. The problems with all such possibilities are socio-economic; such measures are socially disruptive, costly to implement and financially and economically questionable. The economics are especially difficult given the largely low-value crops produced.

To date, the institutional arrangement giving effective control of GWR to the state has not solved the degradation problem. There are two basic difficulties. Firstly, the administration is disparate and has no financial or other incentive to improve water use efficiency, though recently more decision-making responsibility has been passed to farmers. Secondly, the working of the local aquifer, including the way fresh and saline underground water moves and dissipates, is insufficiently understood to facilitate effective management. Imperfect knowledge to a degree thus negates any strategic advantages there may be in centralised management.

The Gaza Strip

The Gaza Strip lies along the south-eastern coast of the Mediterranean Sea and borders the Sinai and Negev deserts. Physically it is not unlike Yao Ba having under 400 mm of rain/yr and largely sandy soils. The Strip has a fast growing population (3.7%) of nearly one million and one of the highest population densities in the world. Over half the population live in refugee camps that are largely unsewered and have limited and often unsafe water supplies. Agriculture has traditionally been the key sector, with citrus the main crop. Fresh water is obtained solely from a coastal aquifer which is now under threat. Historically, the area has been occupied by a succession of powers, most recently by Israel (1967-1994), which have acted against the development of effective management structures.

The Problem

GWR in the Strip are threatened by depletion, contamination, and depletion-induced saline intrusion. Water table levels have dropped drastically since exploitation began early this century, although not consistently throughout the area. Data on contamination is scarce but it is clear that nitrate and probably pesticide concentrations are very high in areas with intensive agriculture and biological contamination is severe in unsewered residential areas. The Strip is surrounded by saline water including the sea and trapped saline deposits (conate water) which are intruding into the aquifer. Areas of freshwater are thus diminishing rapidly and only two areas of good quality water remain.

The Management System

During the Israeli occupation, abstraction was strictly controlled through licensing and quotas but at levels thought to exceed aquifer recharge by 200 to 400%. With autonomous Government the control of abstraction has been relaxed much further. The system today is a prime example of 'open access' exploitation of a common pool resource, with thousands of farmers and households acting rationally as individuals but effectively mining the resource as a whole. The absence of effective Palestinian institutions for water management in Gaza is a basic problem for mitigation.

Strategic Options

As elsewhere, there are three strategic options to aquifer management, to degrade, to manage sustainably, or to restore. As in Yao Ba restoring the aquifer to its previous condition is an unrealistic objective given the prevailing political, institutional and economic circumstances, and appeals only environmentally. However continued degradation is not inevitable, and sustainable management is possible though it may involve accepting the loss of high quality water and using the resource largely for irrigation. Both cases would demand institutional, legal and policy changes of very considerable magnitude, including regulation and control of a large number of individually owned pumps and operators.

Because many individuals would suffer in the process, these changes would give rise to enormous difficulties in implementation and enforcement. New institutional systems and structures, however, are vital if sustainable management is to be taken seriously. At the current rate of degradation, it has been calculated that water of drinking water standard will be available for only some fifteen or twenty years, and water for irrigation water for eighty years. If these projections are correct, at this point there will be no water in the Gaza Strip fit for any form of human activity.

Policy, Theory and Practice

The widespread and chronic underestimation of the value of groundwater systems provides a strong a priori case for the very careful assessment of groundwater management options in the developing world where mistakes can be least afforded. Our case studies have demonstrated the use of an analytical framework for the study of degradation and thrown light on the policy problems facing groundwater management in two contrasting situations. The GWR problems illustrated in both areas are extreme but by no means unusual; water is increasingly scarce and degraded by pollution or salinity.

In both cases the local situation points to the need for a new emphasis on demand management and water efficiency, and balancing economic and equity considerations over space and time. But in both circumstances, the problems of implementation are immense.

In Gaza, the example of open access to the resource raises serious questions about sustainable management and how it could be regulated or enforced. In Yao Ba, local water authorities are strong but unable to apply sensible measures for reducing water consumption due to 'arm twisting' by central Government. At present neither public institutions, local communities nor individual farmers have sufficient incentive to economise on the use of water. The only real disincentive to wastage at present is the recently-increased cost of energy.

While the policy options vary between the two locations, certain of the problems apply generally. One overriding problem in both areas is the paucity of knowledge about the aquifer system and its behaviour, and there is a need for studies of how the aquifer will operate under different management regimes. Such research requires carefully focused and integrated socio-economic and hydrogeologic study aimed at addressing the problems and providing practical solutions. Only such research can answer how different political and institutional systems can address and adapt to the enormous problems facing efficient and sustainable groundwater management in various parts of the world.

7

Groundwater Modelling to Examine Drinking Water Exposure

Prior spatial analyses of two population-based case—control studies of breast cancer in upper Cape Cod, Massachusetts used generalised additive models and geographic information systems (GIS) to map statistically significant areas of increased and decreased odds ratios (ORs). Associations between geographic location and breast cancer diagnosed from 1983 to 1993 were highest when 20 years of latency was assumed. To generate potential hypotheses for further investigation, areas of increased ORs were compared with environmental data available from the Massachusetts GIS portal. A geographic overlap was observed between breast cancer and groundwater plumes contaminated by landfills and wastewater. The boundaries of these groundwater plumes were based primarily on knowledge of area groundwater flow and were developed as a worst-case scenario for contamination around each site.

Based on the locations of public and private wells, we determined that the Barnstable Water Pollution Control Facility (BWPCF) was the only source of wastewater effluent with the potential to impact the drinking water of this study population. There are three public drinking water wells for the Barnstable Water Company (BWC) and several private wells located within the suspected plume boundary. Wastewater is discharged from BWPCF to the groundwater, which flows generally southeasterly toward the ocean. The existence of a sewage plume has been documented by town sources and other research. Groundwater wells in upper Cape Cod draw drinking water from a highly permeable aquifer that is vulnerable to contamination. Concern for the vulnerability of this aquifer has led to recent surveys of local wastewater and groundwater, where suspected endocrine disruptors, including alkylphenols and other estrogenic phenolic compounds, have been detected. Although there is no information on what contaminants may have been in the plumes historically, the low pH and organic carbon content of the Cape Cod aquifer enables many sewage contaminants to persist for up to 30 years in the aquifer.

The geographic relationship between breast cancer clusters and groundwater plumes alone does not establish exposure. The objective of this chapter was to use an extensive exposure model to test the hypothesis that drinking water contaminated by municipal wastewater effluent from the BWPCF may be associated with breast cancer. We used residential histories from 1943 to 1993, questionnaire data on self-reported sources of drinking water (public or private) and on the use of bottled water, operating information for the BWC and BWPCF, and a modified three-dimensional groundwater model from the U.S. Geological Survey (USGS) to estimate when exposure occurred and which participants were exposed. Because the risk of breast cancer may be dependent on the timing of exposure in relation to diagnosis, we examined varying latency periods and exposure durations.

Materials and Methods

Study population. We investigated the association between exposure to drinking water contaminated by effluent from the BWPCF and breast cancer in upper Cape Cod, Massachusetts, using data from two population-based case—control studies: the Upper Cape Cancer Study and Women's Health on Cape Cod Study. The two studies had a combined total of 638 women who had been diagnosed with breast cancer and 842 controls who had been permanent residents of five towns in upper Cape Cod, Massachusetts. We used the Massachusetts Cancer Registry to identify incident breast cancer cases diagnosed between 1983 and 1993. Demographically similar women who also lived in upper Cape Cod were selected as controls by random digit dialing; Medicare records; and records of deceased residents provided by the Massachusetts Bureau of Health, Statistics, Research, and Evaluation. Cases and controls were comparable in age and vital status. We randomly assigned index years to controls in a distribution similar to that of diagnosis years for cases and used the diagnosis and index years to define the relevant time and duration of environmental exposure among cases and controls, respectively. Study participants provided informed consent, and extensive interviews by trained personnel collected information on demographic characteristics and breast cancer risk factors, sources of drinking water (public or private), and residential histories over a 40-year period. Next of kin served as proxies for cases or controls who were deceased or too ill to participate in the interview. All residential addresses in upper Cape Cod were geocoded using GIS. A detailed account of the methods has been reported previously. The Institutional Review Board of Boston University Medical Centre approved this research.

Exposure Assessment

Exposure was assessed using residential histories, questionnaire data, information on the public water supply, and a three-dimensional groundwater model to track the movement of effluent from the BWPCF to the private and public drinking water wells of the study population over time. The BWPCF, operating since 1937, used primary treatment and sand filter beds until around 1980, when secondary treatment was added, before discharging effluent to the groundwater. Approximately 1,5 miles (2.4 km) south of the BWPCF are three public drinking water wells for the BWC, the public water supplier for hundreds of area residences.

The BWC water district has operated since the 1920s, growing from 2 to 12 public drinking water wells between 1920 and 2005; water drawn from all the wells is mixed in common storage tanks (standpipes). Pumping reports from the 1970s and 1980s estimated that these three wells contributed 40-50% of the water supplied by the water company.

The first aim of the exposure assessment was to determine when the drinking water wells were impacted by effluent from the BWPCF. Concerns with the vulnerability of the drinking water prompted the USGS, in cooperation with state and local agencies, to create a detailed groundwater model of the aquifer in upper Cape Cod, an area called the Sagamore Lens. This three-dimensional, finite-difference numerical model solves groundwater flow equations by representing a hydrologic system as a series of discrete spatial units within MODFLOW, a modular software program publicly available from the USGS Web site (USGS 2008).

The model accounts for both horizontal movement of the plume, as well as plume depth in relation to locations of drinking water wells. Uniform geological properties and stresses, such as well pumps, are assigned to each cell of the model grid to simulate steady-state or transient flow conditions. We used MODFLOW-GUI v4 (USGS, Reston, VA), a graphical user interface, with Argus ONE to facilitate our analysis and display results with GIS. Most USGS files were directly imported to this interface, but some files were recreated manually because of software limitations.

The USGS model was created for conditions in 2003, so modifications were necessary to examine historical groundwater movement. We collected data on the history of the public water system from water department personnel, including estimated start dates for wells near the facility. The Hyannisport and Simmons Pond wells came online in 1953 and 1970, respectively, and the Straightway well came online in 1961 and was replaced in 1984. Along with the start of facility operations in 1937 and the last year (1993) in which women with breast cancer were identified, these dates define five distinct time periods with varying well pumping conditions for modelling groundwater flow.

Limited historical data included reports of total withdrawals for all three wells together and pumping rates after the shutdown of the 1961 Straightway well. Pumping rates for 2003 that were used in the USGS model for the Hyannisport and Simmons Pond wells were retained for the historical model. These values seemed reasonable for normal operations over the long period, at 30% and 50% of each pump's maximum rated capacity, respectively, and were plausible given the total withdrawals in the mid-1970s.

The depth and pumping rate for the 1961 Straightway well was unknown, because it was not part of the 2003 USGS model. We included the 1961 Straightway well in our model at the same depth as nearby wells, a likely position indicated by earlier studies. Groundwater flow is highly dependent on pumping rates, so as a sensitivity analysis, three scenarios were modelled for the Straightway well with varying percentages of the rated capacity for the well: 30% capacity (28,875 [ft^3]/ day), 50% capacity (48,125 [ft^3]/day), and 75% capacity (72,188 [ft^3]/day).

Once groundwater flow was calculated in MODFLOW, the results were used by the MODPATH program (version 4.3; USGS) to delineate the path of effluent particles by advection over time to the public and private drinking water wells. Using the same methodology as earlier USGS studies, particle movement in groundwater from the BWPCF was tracked beginning in 1937 to determine if and when those particles reached the wells for each pumping scenario.

Groundwater containing effluent was determined to have reached a well when a particle track ended in the model grid cell containing the well. The exposure assessment assumed effluent moved at the same rate as the groundwater. The study participants served by the BWC were assumed to be equally exposed, because the distribution system stores and draws mixed water from a common standpipe.

Questionnaire data were used to determine the drinking water source (public or private) of the participants for each of their residences. For participants who reported public drinking water, GIS files of the BWC pipe distribution system and geocoded addresses were used to identify which participants were supplied by BWC. Private drinking water wells were assumed to be impacted by the plume if a particle was tracked within a depth of 100 ft (30.5 m) of the residence of the subject. This criterion was used because wells in this type of aquifer in Massachusetts are commonly < 100 ft deep. Participants were also asked if they regularly used bottled water, although this information was not provided by residence. Consequently, bottled water use was examined by stratifying the analysis on whether women ever regularly used bottled water. Lastly, residential histories were used to identify which participants were living at residences during years when drinking water was impacted by effluent. We defined ever exposed as one or more years at a residence supplied by BWC or by an impacted private well. Cumulative duration of exposure was calculated by adding the number of ever-exposed residence years for all addresses.

Data Analysis

The epidemiological analysis first compared ever-exposed women with never-exposed women. Never-exposed women obtained their drinking water from public or private wells in upper Cape Cod that were not impacted by effluent from the BWPCF. We also compared women by duration (years) of exposure to an affected private or public well and stratified by regular bottled water use. Multivariable logistic regression was used to estimate ORs and control for covariates. The OR was estimated using the antilog of the beta coefficient. Statistical stability was evaluated with a 95% confidence interval (CI). Adjusted analyses controlled for age at diagnosis or index year (< 49, 50-59, 60-69, 70-79, [greater than or equal to] 80), vital status at interview, family history of breast cancer based on first-degree relatives, personal history of breast cancer (before current diagnosis or index year), age at first live birth or stillbirth (< 30 years, [greater than or equal to] 30 years, nulliparous), education level (< 12 years, 12 years, or > 12 years), race (white or other), and study population (i.e., study 1 or 2).

Additionally, we adjusted for ever/never exposure to tetrachloroethylene (PCE) from leaching vinyl-lined drinking water pipes, another contaminant in the public water supply for both study populations, and duration of residence in Cape Cod. Positive associations with breast cancer were previously observed for women exposed to PCE-contaminated drinking water and women with longer durations of residence in Cape Cod. The leaching and transport algorithm used to estimate exposure to PCE in drinking water was based on residential history and characteristics of the water distribution system and were described in earlier studies. We also considered a range of latent periods (i.e., 0, 10, 15, and 20 years), because it was possible that the effluent contained pollutants, such alkylphenols and other estrogenic phenolic compounds, that may either initiate or promote breast cancer.

Results

Participants were predominantly white, > 60 years of age, and had > 12 years of education. Most were alive at interview; 30% of cases and 40% of controls were deceased and their information was collected by proxy. More cases than controls had first-degree family history of breast cancer, were greater than or equal to 30 years old at first birth, or were nulliparous.

Groundwater modelling indicated that effluent from the BWPCF reached drinking water wells for the public water supplier BWC as early as 1966. The sensitivity analysis to estimate groundwater flow indicated that effluent would have arrived at different times depending on the pumping rate for the original Straightway well. Assuming a high pumping rate of 75% capacity (72,188 [ft^3]/day), effluent would have reached the original Straightway well in 1966 and the Simmons Pond well in 1973. Assuming a low pumping rate of 30% capacity (28,875 [ft^3]/day), the Simmons Pond well would have been impacted first, and not until 1971, and the replacement Straightway well would have been reached around 1989. The results for 50% capacity were similar to those for 30% capacity.

Because exposure would have begun in 1971 with the low pumping rate and in 1966 with the high pumping rate, there was a 5-year difference in opportunity for exposure. The low and high pumping scenario predicted similar numbers of ever-exposed subjects (n = 241 and n = 247, respectively) when latency was ignored. However, when latency was considered, fewer subjects were classified as exposed when the lower pumping rate was assumed. This finding was mainly because the case ascertainment period was 1983-1993, so when we considered a 20-year latency period, only those participants with a diagnosis or index year during 1991-1993 could have been exposed in the low pumping rate scenario, whereas participants with a diagnosis or index year from 1986 to 1993 may have been exposed in the high pumping rate scenario.

Without considering latency and irrespective of the pumping rate, approximately 78-80% of the subjects were exposed at one address, 17-19% were exposed at two to three addresses, and 3% were exposed at four to six addresses. The overall distribution of exposure duration was similar with each pumping scenario, but the distribution varied over latency periods. Subjects had a cumulative exposure from 1 year to approximately 20-40 years, depending on latency period. Based on the distribution of exposure durations, associations with breast cancer were estimated for > 0 to 5 years, > 5 years, and > 10 years of exposure relative to no exposure at any time during the study period.

When latency was not considered, there were small increases in breast cancer ORs for participants exposed > 10 years compared with subjects who were never exposed [adjusted odds ratio (AOR) = 1.3; 95% CI, 0.9-2.0 for low pumping rate and AOR = 1.3; 95% CI, 0.9-1.9 for high pumping rate]. When a 10-year latency period was considered, larger ORs were seen for the subjects exposed > 5 years (AOR = 1.5; 95% CI, 0.9-2.7 and AOR = 1.4; 95% CI, 0.9-2.2 for low and high pumping rates, respectively). The same was true for exposure > 10 years (AOR = 1.6; 95% CI, 0.8-3.2 for high pumping rate), although AORs could not be calculated for the low pumping rate scenario because of small numbers [crude odds ratio (COR) = 4.6; 95% CI, 1.0-22.3]. Small numbers also limited the adjusted analysis when 15-and 20-year latency periods were considered, but positive associations were seen for ever-exposed versus never-exposed subjects in either pumping scenario (CORs = 1.2-5.9).

The low and high pumping rate scenarios assume different rates for the Straightway well. Particle-tracking analyses indicate effluent would have reached the Simmons Pond well in 1971 for the lower pumping rate and in 1966 for the higher pumping rate. Adjusted analyses controlled for age at diagnosis or index year, vital status at interview, family history of breast cancer, personal history of breast cancer (before current diagnosis or index year), age at first live birth or stillbirth, education, race, and case-control study population. Referent groups for all exposures were women who were unexposed over the entire study period. In the low pumping scenario, 704 controls and 535 cases were unexposed. In the high pumping scenario, 700 controls and 533 cases were unexposed.

A total of 342 participants reported ever regularly drinking bottled water, and 1,138 reported not doing so. Across all latency periods and pumping rates, we found statistically significant associations for ever-exposed versus never-exposed participants among women who did not regularly use bottled water and nonstatistically significant inverse associations among women who did regularly use bottled water. The strongest association was observed for women who did not use bottled water when 20 years' latency and a high pumping rate were assumed (AOR = 2.5; 95% CI, 1.2-5.3 and AOR = 1.0; 95% CI, 0.3-3.3 for nonbottled water users and bottled water users, respectively).

In this study, we developed a historical groundwater model and examined participants' residential histories and drinking water source to determine exposure to drinking water contaminated by effluent from the BWPCF. Recent space-time analyses showed increased breast cancer ORs among women who lived near the BWPCF compared with women who lived in other parts of the study area during the 1960s. We used an existing groundwater model and particle-tracking analysis to estimate the movement of effluent discharge from the BWPC facility to the drinking water wells. Although several assumptions were required to address historical conditions, the model provided a method to explore the spatial and temporal relationship between a source of contamination and a possible exposure route for study participants. We found increased breast cancer for subjects ever exposed to contaminated drinking water wells when considering various latency periods, but little to no evidence of an association when latency was ignored.

The BWPCF was and continues to be a major source of groundwater recharge in upper Cape Cod. Evidence of this is provided by drinking water quality measurements of nitrate-nitrogen (nitrate-N) collected since the Safe Drinking Water Act was promulgated in 1972. Nitrate-N is an indicator of contamination from septic systems or wastewater recharge that may include potentially harmful contaminants such as synthetic organic chemicals. Nitrate-N concentrations in drinking water wells in Cape Cod < 0.5 mg/L are considered background levels, and concentrations > 1.0 mg/L are due to human impact, typically from wastewater or fertilizer use. The U.S. Environmental Protection Agency maximum contaminant level for nitrate-N is currently 10 mg/L. With the exception of a few years, the nitrate-N levels at the public drinking water wells near the BWPCF were > 1.0 mg/L and often approached 5 mg/L from the early 1970s to 1990s.

The overall trend in the monitoring data shows that nitrate-N measurements increased at a similar rate as the quantity of discharged effluent over time. This suggests that effluent impacted the monitoring wells, although increases in residential septic use may also have contributed to these increases. Because historic monitoring was only of nitrate-N, we do not know what chemicals may have been present in the effluent when contamination of the drinking water wells first occurred in the late 1960s and early 1970s.

Results from our groundwater model were comparable to earlier studies with regard to plume distance and depth, but differed in direction. In the mid-1980s, studies documented that the sewage plume extended 5,500 ft (1,646.4 m) south of the facility, with a depth of 20-65 ft (7.6-19.8 m) below the surface bounded vertically by an impermeable clay deposit. The depth of the plume was within the range of the BWC wells between 61 and 75 ft (18.6-22.9 m) deep, although there was some evidence that the deepest well, the Simmons Pond well, may have penetrated a lower aquifer system and thus was protected from contamination above by the layer of clay. However, the earlier studies concluded that several ponds in the area directed the wastewater plume toward the harbour, completely by-passing the public drinking water wells. The surface water interactions that we incorporated into our historical groundwater model indicates this may have protected the Hyannisport well, but the Straightway and Simmons Pond wells may still have been affected. Our model found an overall larger area impacted by groundwater originating at the facility than was delineated by these earlier plume studies and thus included the drinking water wells. The model also found it took approximately 30 years for the effluent to reach the public drinking water wells. A sewage plume in the nearby town of Falmouth showed similar persistence and behaviour under similar geologic conditions.

In the 1980s, low levels of other wastewater indicators (specific conductance, ammonium, alkalinity, and potassium) at monitoring wells 2,000 ft (609.6 m) north of the drinking water wells were cited as evidence that the plume was not moving in the direction of public drinking water wells but was passing southeasterly to the ocean. However, some uncertainty surrounds this conclusion because these samples did not represent historical conditions or conditions at the drinking water wells located to the south. At the time of the studies in the 1980s, additional monitoring was recommended to establish the source of the elevated nitrate-N measurements, as the existing monitoring wells were considered to be insufficient. It is possible that monitoring in earlier studies did not adequately delineate the edges of the Barnstable plume and its southerly extent, so drinking water wells may have been impacted by the sewage plume in prior decades as our model results suggest.

To the best of our knowledge, only one other epidemiologic study has investigated the association between breast cancer and exposure to drinking water potentially contaminated by wastewater that study, which was also conducted among women living in Cape Cod, used nitrate-N levels measured in public water supply wells beginning in 1972 as a measure of exposure to public drinking water contaminated by onsite wastewater disposal (septic systems). The study did not find evidence of an association with breast cancer after considering a range of latent periods.

Differences in study methods, including exposure definitions, may account for the dissimilar findings. In the Brody et al. study (2006), cancer cases were diagnosed between 1988 and 1995 while living in one of the 15 Cape Cod towns. Because the exposure data only went back to 1972, participants were included in the study only if they had lived in Cape Cod on a public drinking water supply during the 16-year period prior to diagnosis (or index year for controls). In contrast, our study examined a smaller region of Cape Cod using cases diagnosed between 1983 and 1993. We also included both public and private well users and modelled exposure beginning in 1966. The main difference is that the Brody et al. study assessed the widespread exposure to public drinking water impacted by residential septic systems, whereas the current analysis assessed exposure to effluent from the BWPCF. The Barnstable treatment facility is a considerably larger point source, concentrating waste from across a broad geographic area and combining both residential and commercial wastewater.

This study investigated a hypothesis generated by prior spatial analyses. An important strength of this analysis was the use of various data sources to reconstruct historical exposures with respect to both space and time. We integrated groundwater modelling, residential mobility, and information about public water systems in CIS to assess exposure to drinking water impacted by wastewater effluent. Despite out best efforts in modelling groundwater movement and determining who was exposed, the analysis is still limited by exposure misclassification. Assumptions concerning operations of the public water system, wastewater facility, and groundwater conditions were necessary. Our sensitivity analysis showed that the exposure period depended on pumping rate conditions for a public well with missing information. In addition, unknown changes over time in drinking water withdrawals, aquifer conditions (i.e., precipitation, impact of facility effluent discharge on groundwater levels), distribution system infrastructure (adding pipes, water storage, mixing), and personal water usage also affected the exposure of subjects at their residences.

The misclassification of exposure status, however, was likely nondifferential, as both cases and controls in this area were served by the same supply. Another limitation is that the composition of the wastewater effluent was unknown, other than some compounds such as nitrate-N after 1972. Thus, we could not account for reactions within the plume and aquifer that may have slowed the movement of contaminants through the aquifer.

In addition, we could not account for changing concentrations over time. However, we conducted an additional analysis using nitrate-N concentrations at the drinking water wells as an indicator of trends in the concentrations of other potential contaminants. We created a scaled relative measure of exposure by adding available nitrate-N concentrations at the three Barnstable wells over the residence period for exposed subjects (SSI 2008).

For a 10-year latency with low pumping rate scenario, the AOR was 1.7 (95% CI, 0.9-3.4) for the highest nitrate-N exposure quartile (versus no exposure), which is similar to the association with > 5 years of exposure (versus no exposure) to the contaminated plume (AOR = 1.5; 95% CI, 0.9-2.7). We found the same association with this nitrate-N measure across latency periods. We feel these results support our original assumption that exposure is driven by duration relative to the timing of effluent reaching the well.

The current study results may also have been affected by uncontrolled confounding by other environmental exposures. For example, the first Straightway well was closed because of contamination by volatile organic compounds, with little further explanation. In addition, the 1982 Upper Cape Cancer Incidence Study, which examined cancer in relation to many different environmental exposures, found it difficult to separate exposure to the BWC and the Barnstable Airport.

Most individuals served by the BWC also live in proximity to the airport. That study did not find an association between breast cancer and contaminated drinking water from BWC, but the case ascertainment period was 1983-1986 and few participants-were exposed for long durations. There may also have been uncontrolled confounding from pesticide exposure from cranberry bog cultivation.

We did address PCE contamination and residency duration as confounders in our analysis, but they did not affect the results and were not included in the final model. We also estimated associations with breast cancer after excluding the 86 women with a prior history of breast cancer and observed similar point estimates and CIs. It is unlikely that the results were biased by missing data, as residential mobility and water type were reported with similar completeness for cases and controls. In addition, associations with established breast cancer risk factors were present in the study populations, as previously described.

When cancer clusters are discovered, there are many possible environmental factors that could be investigated. In this study, given the geographic overlap of groundwater plumes identified using spatial analyses, we investigated the hypothesis that exposure to drinking water contaminated by effluent from a wastewater treatment facility was associated with breast cancer. The spatial relationship alone does not establish exposure, but the current study was able to determine a plausible route of exposure by also taking time into account. Assessment of historical environmental exposures often requires substantial effort, and the general methodology described here can be applied elsewhere. Although definitive associations were not found in this study, we showed that by incorporating additional data, such as residential histories of the participants and contaminant movement over time, hypotheses generated by spatial analyses can provide additional insights into the environmental etiology of breast cancer.

This analysis found evidence of a positive association between breast cancer and exposure to drinking water impacted by wastewater effluent from the BWPCF. The associations were strongest among women who were not regular bottled water users and among women exposed for long durations when latency periods were taken into account. The current exposure analysis expands on our earlier work to explore the spatial and temporal relationship between a source of environmental contamination and a route of exposure for this study population.

Our prior spatial analysis identified groundwater plumes as a potential environmental exposure that had not yet been fully studied. We investigated this hypothesis using a detailed groundwater model and determined that contamination of drinking water by effluent from the BWPCF was plausible. Area groundwater sources for drinking water are subject to more protections now, and the impact of sewage on groundwater was carefully considered in recent expansion plans for the facility. However, this analysis suggests the sewage plume emanating from the facility may have had a significant historical impact on drinking water.

Management of Mexican Effluent

The cities of Nogales, Arizona, and Nogales, Sonora (together known as Ambos Nogales) have many cultural and historic bonds that tie them together. The communities also share surface and groundwater resources, which have been the source of both cooperation and conflict. Population pressures have placed added demand on the shared resources and have caused the communities to search for more water sources to supplement traditional supplies. In addition, wastewater from both cities is treated nine miles north of the border in Rio Rico, Arizona, at the Nogales International Wastewater Treatment Plant (NIWTP). Ambos Nogales is located in a semiarid region with high climate variability characterised by frequent drought and occasional serious flooding.

The prominent water resource in the area, supplying both Nogales, Arizona, and Nogales, Sonora, is the Santa Cruz River. With its headwaters in the San Rafael Valley, the river flows southward through Sonora, Mexico, for twenty-five miles and returns to Arizona about five miles east of Nogales, Arizona (ADWR 1999). River flows are ephemeral or intermittent with some perennial reaches. A perennial reach of the river exists downstream of the NIWTP, caused by effluent discharge into the river from the plant. On both sides of the border, the Santa Cruz River aquifer system is generally shallow with limited storage capacity and extreme sensitivity to drought. The aquifers also recharge quickly when rain is present. The Mexican effluent is an important water resource in times of drought because of the low storage capacity of the river aquifers.

Effluent is an increasingly important renewable water resource in Ambos Nogales. Two-thirds of the effluent treated at the NIWTP is wastewater generated in Mexico. The treated water flows northward, revitalizing the otherwise dry Santa Cruz River, recharging water tables, and supporting a rich riparian habitat downstream of the plant. International Boundary and Water Commission (IBWC) Minute 294 established Nogales, Sonora, and Nogales, Arizona, treatment capacity at the NIWTP at 9.9 million gallons per day (mgd) and 7.3 mgd, respectively. Mexican wastewater flow to the NIWTP has averaged 9.6 mgd in recent years, yet Mexico exceeded its allotted capacity in 2000 and 2001.

While U.S. water-rights holders and environmentalists along the border want continued use of the effluent, Mexico retains legal right to it. At some point, Mexico could decide to use the effluent for industrial, agricultural, or aquifer-recharge purposes within its own boundaries. In this chapter I examine the issues related to the management of Mexican effluent in Ambos Nogales. I describe possible mechanisms for management and how new developments may present opportunities for an agreement between Mexico and the United States. I present three scenarios for management of the Mexican effluent that would address the needs of both countries.

Border Effluent

In a broader context, the southwestern United States and northwestern Mexico have a generally dry climate and rapidly growing populations. Since both countries use the same limited water resources, the use of effluent is one way to increase the quantity of available water for both countries. In Ambos Nogales, one country (Mexico) generates wastewater up-gradient that is treated and used in another country (Arizona) down-gradient. Although there are concerns related to the effluent's effect on groundwater quality, the effluent has become a vital part of the water budget for southern Arizona and a potentially important source for Nogales, Sonora. Without the Mexican effluent in the water budget, the demand for water in the Nogales, Arizona, area would exceed supply by some 7,000 acre-feet per year.

Persons proposing to offer subdivided lands for sale within an Active Management Area (AMA) must demonstrate that sufficient water of adequate quality will be continuously available to meet the water needs for the proposed use for at least one hundred years. In the Santa Cruz Valley, the problem lies in the legal complexities that surround the effluent in both Mexico and Arizona. While the effluent currently is discharged and utilised exclusively in Arizona, Arizona law (A.R.S. [section] 45-576) restricts its use because it belongs to Mexico, making its long-term availability uncertain. IBWC Minute 276 (IBWC 1988) gives Mexico the legal fight to use its portion of the effluent, even though it is treated in Arizona. Yet because of the geography of the area, it would be expensive to pump the effluent back uphill to the city of Nogales, Sonora. Mexico is in the position of both losing its water and incurring a cost for treating the water ($200,000 per year; Steve Tencza, personal communication, September 19. 2002). Presently, no group or agency in Arizona is either willing or able to compensate Mexico for its effluent. In a nutshell, Mexico owns the effluent but doesn't have control of it, and Arizona has control of the effluent but doesn't own it.

Negative environmental consequences in Arizona could result if Mexico decides either to recapture its effluent or not allow its wastewater to cross the border. Either action would dry up much of the ten-mile reach of riparian area that is dependent upon Mexican effluent for its existence. The Gila topminnow, a federally listed endangered species, resides in the reach of the effluent-dominated Santa Cruz River downstream of the plant, as does the endangered southwestern willow flycatcher.

The political effects of resolving the effluent issue would be better relations between the two border communities and decreased tension over a lingering issue that is in contrast to the generally cooperative attitude that characterises relations between Nogales, Arizona, and Nogales, Sonora. The issue has created concern on the Mexican side that Arizona would use the Mexican effluent and eventually claim possession of it. The apprehension in Arizona is that Mexico will unilaterally decide to take back its effluent, which would negatively affect riparian habitat, aquifer recharge, and other indirect uses that rely on the effluent. The average water consumption per capita is four times higher in Nogales, Arizona, than in Nogales, Sonora, which illustrates the social and economic disparities between the two countries.

This discrepancy results in feelings in Mexico that Arizona uses the effluent for riparian areas and for new housing, where Mexico could potentially utilise it to meet the more basic needs for drinking water for its existing populace or directly for industrial processes and irrigation. While Mexico does not imminently threaten to take back its wastewater, an overt, unilateral action on the Arizona side to use the Mexican effluent could spark Mexico to act. Mexico has a right to stop the wastewater from crossing the border or to capture the effluent and return it to Mexico. Either action would be both prohibitively expensive or a logistical nightmare, but countries don't always act rationally when they feel they have been wronged.

Possible Mechanisms for Water Distribution

With no immediate crisis looming, now might be the time to examine some possible mechanisms of border water distribution that have been proposed in the past or are planned for the future. One is the Bellagio Draft Treaty, designed for the binational management of border water. While written to address groundwater issues, one section, which dealt specifically with the equitable apportionment of both surface water and groundwater, could be applied to the issues in Ambos Nogales.

The authors of the Bellagio Draft Treaty envisioned a strong binational border agency with the authority to intensively manage water resources in ways not previously done. However, these mechanisms have yet to be put into practice, and no binational agency has emerged that could comprehensively manage border water. The logical choice for such an agency would be the International and Boundary Commission, which is already involved in boundary water management activities, yet it has never exercised the authority Hayton and Utton envisioned.

Consejos binacionales (binational councils) are a second, more recent contribution to regional water management that incorporate equity, concerns with regional water policy. Mexico has developed consejos de cuencas, or watershed councils, to manage and preserve water resources in specific regions that require close management.

Brown and Mumme cite an example of a nascent version of such a watershed council in the San Diego—Tijuana area that is operating fairly efficiently. While no such council presently exists in Ambos Nogales, as competition for water increases, this would be a very useful alternative.

The regional workgroups proposed in the Environmental Protection Agency's (EPA) Border 2012 Program is a similar approach to watershed councils, which could help to strengthen cross border communication and coordination. The EPA considers regional watersheds the "foundation" of a new program that would emphasize a bottom-up approach, with strong local input, focused on achieving concrete, measurable results. We should be able to evaluate in a year or two how effective these groups will be in addressing border water issues.

Water Management Authority

While the Santa Cruz AMA is responsible for managing water resources in southern Arizona, it lacks the tools to meet the stated AMA goal: "to maintain a safe yield condition and to prevent local water tables from experiencing long-term declines". Some issues affecting management of the AMA include the existence of more legal claims to Santa Cruz River water than there is actual water; the AMA has authority to manage only groundwater, not surface waters, and the AMA has no powers to physically move water from sources of supply to sources of demand (ADWR 2000). In order to manage water resources in southern Arizona more comprehensively, the establishment of a water management and importation authority has been proposed by local water interests.

The authority would be an entity that could comprehensively manage wastewater and potable water for the upper Santa Cruz watershed in southern Arizona. Such an authority could acquire and provide water to its members, recharge and store water, and build infrastructure to move water around (State of Arizona 2002). While it remains to be seen what relationship would exist between the Santa Cruz AMA and the water management authority, ideally they would work collaboratively, with the authority working to manage water (e.g., surface water and effluent) in areas where the legal power of the AMA does not reach.

The water management authority could also be the vehicle to negotiate with Mexico about continuing wastewater flows to Arizona. The key to making an agreement work would be to ensure that both Nogales, Sonora, and Nogales, Arizona, have their respective water needs met (Sprouse and Vaughn 2002). An equitable agreement between the two countries on effluent could be a link that binds the two communities instead of being a point of contention.

Alternative Effluent Management Scenarios

A water management authority could choose various strategies to meet its goals. I examine three water management scenarios, each with a different strategy to manage the Mexican effluent. In whatever strategy is adopted, effluent-use alternatives would have to be mutually beneficial and address concerns of stakeholders on both sides of the border. The alternatives begin with the simplest solution to implement (Scenario 1) and progress to the most complex (Scenario 3).

Scenario 1: Diversion of Effluent to Municipal Water Providers

Arizona would pay Mexico fir the treatment costs of its effluent, and Mexico would give up its legal right to the water. The water management authority could be the entity that actually pays for the treatment cost of the Mexican effluent and is then, in turn, able to satisfy the Arizona Department of Water Resources' (ADWR) requirement of a hundred-year supply of water for new housing development within the Santa Cruz AMA. The authority could create a cash flow by charging municipal water providers in Nogales, Arizona, Rio Rico, Tubac, and Amado for the water. Nogales, Arizona, would likely use its apportionment of effluent to recharge the Guevavi municipal well field. Revenue collected from new homeowners utilising Mexican effluent would pay for the treatment costs. A base flow of approximately 5,000 mgd would need to be left in the Santa Cruz River in order to maintain riparian habitat.

This alternative could require participation by the U.S. government (Meyers 2000). As new housing subdivisions are built in the area, the authority would be gradually able to generate a cash flow to Mexico over a twenty- to thirty-year period, up to the point where it would pay the full cost of treating the Mexican wastewater. During this twenty- to thirty-year period, the U.S. government would pay the difference between the authority's contribution and the cost of treating the effluent (or the amount charged by Mexico). If Mexico agreed only to twenty- or fifty-year contracts to deliver effluent, ADWR rules would have to be modified to accommodate the Mexican terms. The advantages of this scenario are that it is the least complicated, would be the easiest to implement, and provides water to the population centres where it is most needed. The disadvantage is that it diverts water which would otherwise stay in the Santa Cruz River channel.

Scenario 2: Provision of Effluent to Downstream Users

This option differs from Scenario 1 in that the Mexican effluent would be sold to downstream users located at the northern end of the Santa Cruz AMA or to entities in the Tucson AMA. This is an area with large aquifer storage capacity and significant amounts of developable land, particularly in the southern Tucson AMA. The advantage of this option is that flowing water would be retained in the Santa Cruz River for preservation of riparian habitat while supporting new growth to the north. The water would be used only after it has passed through, and recharged, the riparian areas located between Rio Rico and Tubac. The advantage of Scenario 2 is that water remains in the river for preservation of riparian habitat. The disadvantage is that the largest water providers, Nogales and Rio Rico, don't get the water. Both communities are located upstream (south) of the riparian areas, and if they were to utilise the effluent, this would result in reduced flows of water to support riparian habit.

Scenario 3: Return of a Portion of the Effluent to Mexico

Under this scenario, some of the effluent originating in Mexico would be returned to Mexico. The population of Nogales, Sonora, continues to experience significant growth, with substantial increases in demand for potable water and sewerage collection facilities. This scenario makes a more realistic assumption regarding Mexico's attitude toward its water: that Mexico views its water as a part of its patrimony, as it did oil in the 1920s, and that the value placed on water exceeds the cost in dollars of treating the water. Initially, Mexican effluent in excess of 9.9 mgd could be sent back to Mexico, with the remainder of the Mexican effluent used either for municipal water providers or for downstream users in Nogales, Arizona.

Returning water to Mexico could be done in one of two ways. One way would be to transfer a portion of the effluent from the NIWTP to the Mascarenas well field on the Santa Cruz River, located just south of the U.S.-Mexican border.

This would allow the effluent to be recharged into the aquifer, taking advantage of soil filtration to remove any impurities still in the water, and then pumped back for use as drinking water in Nogales, Sonora. This alternative has the advantage of creating an almost closed conveyance system where the water continuously cycles from Mexico to the NIWTP and back to Mexico. To help lower costs, this project could be undertaken in conjunction with the city of Nogales, Arizona, which has proposed using its effluent to recharge the Guevavi well field along the Santa Cruz River.

An extension to the city's effluent pipeline could carry the water the additional five or six miles to the border. By itself, an effluent pipeline from the NIWTP to the border would not be cost effective, however.

A second way that water might be returned to Mexico would be to provide potable water directly from the city of Nogales, Arizona, to the city of Nogales, Sonora. Presently, there are three small potable-water pipelines by which the city of Nogales, Arizona, provides water to customers in Nogales, Sonora. One connection goes to a hotel, while the passage of time has created uncertainty as to where the other two connections go. The water authority could increase the size of the pipelines to supply more potable water to Nogales, Sonora, in exchange for continued flows of Mexican effluent into Arizona. This plan would require that more effluent be recharged to one of the well fields in Arizona, in order to meet whatever additional supply is given to Mexico. Once again, the plan has the advantage of allowing the water to cycle to and from Mexico.

In both versions of this scenario, Arizona's needs are met by having guaranteed effluent flows from Mexico, and Mexico's needs for potable water are met by the returned water. This alternative would allow Nogales, Sonora, to reduce the amount of water being pumped from the Los Alisos basin to the south, whence it must be pumped over a basin divide to Nogales, Sonora.

The water in Los Alisos could be used to support projected growth in that area while the needs of Nogales could be provided for by the effluent from the NIWTP (or by potable water pipelines). The water authority would pay for treatment costs of Mexican wastewater at the NIWTP and would share costs for building a pipeline and pumping effluent back to Mexico.

The advantage of this scenario for Arizona is that it would receive a hundred-year guarantee of flows from Mexico. Mexico would receive (1) cancellation of a yearly debt to pay for treatment of its wastewater in Arizona, and (2) a portion of its effluent back to satisfy demands of its growing population. The disadvantages are the cost of building infrastructure to deliver effluent to Mexico and the operation and maintenance costs.

Table: *Population figures for Nogales, Arizona, and Nogales, Sonora*

Year	*1950*	*1970*	*1990*	*2000*	*2010*
Nogales, Arizona	6,153	8,946	19,489	20,878	24,282
Nogales, Sonora	26,016	53,494	107,936	206,554	275,704

Sources: IBWC 1998; U.S. Census Bureau 2000.

Legislation that would establish a water management authority in Santa Cruz County is projected to be submitted to the Arizona Legislature in the 2003 session (Alejandro Barcenas, personal communication, October 23, 2002).

This would be the first step toward equitable management of Mexican effluent in Ambos Nogales. While it would require extensive negotiations on both sides to overcome political obstacles, many other pieces are in place to open a window of opportunity and make an agreement possible:

- The water rights along the Santa Cruz River are going through a settlement process that would allow a water management authority to identify who owns water rights and who would have to pay to use Mexican effluent.
- The current U.S. and Mexican presidents are able to work together and have an interest in border issues.
- An economic recession makes income from the effluent more important to Mexico than in the past.
- Groups in southern Arizona are willing to work together.

The Home Sewage Disposal Systems

Water is arguably one of our most valuable resources, yet the quality of this resource remains a serious issue facing both the federal government and the state of Ohio, because these entities are challenged with the task of its protection. While we often think of waterborne diseases from contaminated drinking water as being a problem for poorer countries or third world locations, from January 2003 to December 2004, 30 reported outbreaks occurred of waterborne diseases associated with contaminated drinking water in the U.S.

A critical part of protecting water quality is the appropriate collection, treatment, and disposal of domestic wastewater. Within the U.S., while most domestic wastewater management is a centralised system (i.e., collection into sanitary sewers and treatment at a wastewater treatment plant that typically produces secondary or tertiary treated effluent), about 25% of the country's population and 33% of new developments are served by decentralised, often individual onsite wastewater treatment systems. The U.S. EPA has found that such on-site treatment systems offer cost-effective alternatives for waste treatment and the protection of human health.

Ohio does not vary from the national trend, with approximately 25% of Ohio's households being served by some type of onsite sewage treatment systems. From 1979 to 2000, the most commonly installed system in Ohio was the septic (anaerobic) system with a soil adsorption field for ultimate treatment and disposal, accounting for approximately 75% of the total on-site, individual wastewater treatment systems (CT Consultants [CTC], 2001). This trend seems to be holding steady or slightly increasing according to the Ohio Department of Health (ODH), which reported that between January 1 and June 30, 2007, 87% of systems installed were septic tanks discharging to soil absorption systems.

Onsite systems can also contribute to contamination of drinking water sources, however. U.S. EPA has estimated that as many as 168,000 viral illnesses and 34,000 bacterial illnesses occur each year from contaminated drinking water and that onsite sewage treatment systems (STS) are one potential source of the contamination. Yet, despite the extensive use of onsite STS in Ohio, little knowledge currently exists about the actual failure rate of these systems; some states report failure rates as high as 50% to 70%. The only comprehensive survey of systems in Ohio reported in the literature was performed by Mancl (1990). That study, however, did not examine systems in the field, but rather surveyed county health departments to estimate failing systems, which was reported to be approximately 27% statewide. In this chapter, we report on failure rates in onsite STS that were found as part of a comprehensive seven-county survey that was performed under the auspices of the Northeast Ohio Areawide Coordinating Agency (NOACA) during the summer of 2000. A complete report on all aspects of the project, which included development of extensive databases, can be found in the NOACA Report (CTC, 2001). This chapter focuses solely on the information collected on failing systems as part of the comprehensive report.

Definition of Failure

The ultimate goal of a home sewage disposal system (HSDS) is to treat and dispose of home wastewater by infiltration into a soil system. Typically, design criteria implemented by the counties through regulation are designed to ensure that systems are installed significantly above the water table to avoid groundwater contamination. While many factors can contribute to HSDS failure, soil unsuitability is often a primary cause. Unfortunately, the majority of soil types in the northeast Ohio are classified as being severely limited for subsurface disposal of effluent by the USDA Soil Conservation Service. For example, out of 60 different soil types found in Geauga County, 95.0% are severely limited based on soil type, degree of slope, and depth to bedrock. Only 3.3% of the soil types are rated as being moderately limited, while the remaining 1.7% are classified as being only slightly limited. Hence, the issue of most concern to county health officials is failure of these systems through surfacing of wastewater. The goal of the study reported here, then, was to determine the percentage of systems in the seven-county area that were "failing" with respect to the elimination of the direct introduction of wastewater into surface waters.

In reality, failure of the system can occur in one of two modes. Either the system can fail to filter and biologically degrade the waste before it reaches the groundwater table, or the soil system becomes overloaded and wastewater rises to the surface. Because it was prohibitively expensive and practically impossible to drill wells, collect soil and groundwater samples, and analyse for various wastewater constituents (e.g., nutrients, coliform bacteria) for the large number of systems across the seven counties involved in our survey, a system was identified as "failing" if, upon inspection, it had observable surfacing of effluent from the treatment system. For analyses, a system was identified as "surfacing" if surfacing effluent was seen.

Sampling Approach

Grouping of On-Lot System Types and Controlling Variables

Because numerous system types and several potential controlling variables exist, it was not possible to develop a sampling scheme with 490 "samples" (number of on-lot system inspections initially planned to be performed) that would allow statistically and scientifically valid analyses of all potential contributors to system malfunction. Therefore, based on the type of systems actually sampled, and after discussions with the County Health Department representatives, NOACA, and their consultant, we decided to focus on the effect of two control variables: oxygen status (aerobic vs. septic) and filtration (filter vs. no filter). The systems were grouped into four sets based on the two control variables.

Field Survey

A certified soil scientist accompanied by a representative of the relevant local health district conducted each on-site investigation for this study. By having the same soil scientist do all surveys, consistency in methodology of assessing the site and determining if the system was surfacing were ensured throughout the seven-county area. In addition, the soil scientist was able to provide verification of soil types for each installation.

Confidence Intervals

During the investigation, on-lot systems were recorded as either surfacing or not surfacing. Similarly, off-lot systems were recorded as having poor effluent or not based on odour and colour. To calculate a confidence interval, these "yes-no" answers were converted to 0s and 1s. It does not matter statistically what numerical values are used for "yes" and "no" because the statistical method normalises for this selection. Standard practice, however, is to use 0 for "yes" and 1 for "no." This protocol was followed throughout the analyses in this report. Both the 95% and 90% confidence intervals are discussed below.

Significant Difference in Means

The sampling rationale and system grouping built into the survey design allowed comparative analyses to be conducted on the data collected from the inspected systems, as well as grouping system types to allow more robust statistical analyses. The grouping of the systems was designed to allow the analysis of the effect of aeration and filtration on system performance. The percentage of systems failing (i.e., surfacing) was compared across groupings using a standard t-distribution and comparison of the means. Difference is significant when

$$d > (ta/2)\ (s) = (1/na + 1/nb)1/2 \qquad (1)$$

where

d = absolute value of the difference between the means;

(ta/2) = t-value at desired confidence interval (90% used);

na and nb = number of samples in populations a and b, respectively; and

s = pooled standard error, calculated using Equation 2 (below) and

$$s = \{[(na - 1)sa2 + (nb - 1)sb2]/\ [(na - 1) + (nb - 1)]\}1/2. \qquad (2)$$

It provides the total percentage of surfacing systems and the failure rates by control variable (system type and the 90% and 95% confidence intervals for each value). From the data, one can be 95% sure that at least 12.7% and as much as 19.7% (16.2% [+ or -] 3.5%) of all on-lot systems in the seven-county survey area met the definition of a surfacing system as defined in this study. The data does not show any significant differences in the percentages of surfacing systems between any of the various groupings possible. This indicates that the addition or exclusion of either an aeration system or a filter to a home sewage disposal system does not significantly impact the probability of the system failing due to surfacing.

Effect of Soil Rating

The inspection team also verified soil classification and rating at the location of the on-lot systems inspected. Because soil type is considered to be a significant factor in on-lot system performance, it is of great interest to examine the differences in the percentages of systems surfacing in different soil types. Soil ratings and classifications were based on the USDA county surveys of soil types. It is important to note that the soil ratings and classifications found in the records at the various counties did not correspond completely with the reality found in the field. All analyses in this study were performed on soil ratings and soil classifications as verified during the inspections.

The data indicates that systems installed in severe soils have a significantly higher percentage of systems that were found to have surfacing effluent (23.3%) than the overall percentage of surfacing systems (16.2%) at the 95% confidence level. Even with a limited data set for non-severe soils, the data show that systems in "slightly inadequate" soils show a significantly lower percentage of surfacing systems (2.8%) at the 90% confidence level.

Because the severe soil classification represents 63% of the systems investigated, additional information on the parameters leading to a severe soil rating were also analysed.

A soil is rated as "severe" in the soil classification handbook for six reasons: a) excessive wetness (categorised as "wet"); b) slow permeability ("slow"); c) shallow bedrock ("shallow"); d) inappropriate slope; e) susceptibility to flooding; and f) susceptibility to ponding. Sufficient data existed to show soils that were rated as severe because they were wet, wet and slow, or wet, slow, and shallow with respect to the impact on the percentage of systems found to be surfacing. The data indicates that systems in soils that have any combination of excessive wetness, poor permeability, or shallow bedrock have a significantly higher probability of surfacing than systems in other soils rated "severe" by inspection (e.g., inappropriate slope, susceptibility to flooding, and susceptibility to ponding).

Effect of Soil Classification

Only those soil classifications in which at least 10 or more systems were investigated and in which the variance of the control variable (presence of surfacing effluent) was sufficiently low could be used to calculate reliable statistics on the percentage of surfacing systems. For five soils, sufficient data was collected to calculate reliable statistics. The data reveals that those systems in Chili soils or soil complexes have a significantly lower percentage of failing systems at the 95% confidence interval. Out of 48 systems inspected in this soil type, none was reported as surfacing. Conversely, systems installed in either Mahoning or Ravenna soils (or soil complexes predominantly of these types) have a significantly higher percentage of trouble. Systems installed in Wadsworth soil or soil complexes have a significantly higher percentage of surfacing systems at the 90% confidence interval, but are not significantly different from the overall percentage at the 95% confidence level.

Effect of Inspection Date

Because numerous systems were inspected, a time factor existed in the inspection scheme. In addition, different weather patterns before an inspection (e.g., significant precipitation or lack thereof) could also have had an impact on the determination of whether or not effluent was surfacing in a given system. To examine the effects of the date of inspection, the systems were grouped by the month inspected, and the percentage of surfacing systems compared.

An analysis of the percentage of surfacing systems found during each month indicates that little effect occurred in the rate at which systems were identified as "surfacing" depending on the month in which the system was inspected. Only systems inspected in May were more likely to be identified as surfacing, and then only at the 90% confidence level. Because only about 5% of the total number of systems inspected was inspected in May, this difference does not represent an overall significant effect on the data collected.

Correlations between Selected Disposal System/Site Characteristics and Surfacing Effluent

During the investigation of on-lot systems, information on the number of bedrooms, tank volume, and other variables was collected. This data was used to determine whether systems that were found to be surfacing were more often associated with characteristics of the homes (e.g., number of bedrooms) or system design characteristics (e.g., tank size filter bed area, surface slope, and depth to ground water as indicated in as-builts). Sample correlation coefficients were calculated between comparison sets of data were calculated using Equation 3 as follows.

R([y1],[y2]) = S[([yi]-[yave1])([y2]-[yave2])]/(n-1)/[s1][s2] (3)

where

R([y1],[y2]) = the sample correlation coefficient between variable [y1] and [y2];

[yave1] = average of variable [y1];

[yave2] = average of variable [y2];

n = number of sample pairs; and

[s1] and [s2] = standard deviation for variable 1 and 2, respectively (calculated by Equation 2).

The coefficient ranges between 1 and -1 and indicates if the two variables are more similar than would be expected if the two variables were completely independent. A sample correlation coefficient of 1 means the parameters is directly correlated and a coefficient of -1 indicates an inverse relationship. The closer the correlation coefficient is to either 1 or -1, the stronger the correlation between the two variables. It is important to understand, however, that correlation does not indicate causation but merely that the two variables appear related in the data. This study represents a comprehensive survey of onsite home wastewater treatment systems in Northeast Ohio. The survey revealed several important facts about this method of wastewater treatment and disposal in the region. Most importantly, the survey revealed that at least one in eight (12.7% and up to as many as one in five [19.7%]) of the onsite wastewater treatment systems are failing in that they are allowing wastewater to surface as opposed to infiltrate. The rate of failure of these systems does not vary significantly between aerobic and septic systems or between systems with or without filters. Systems installed in soils rated as "severe" are significantly more likely to fail than those installed in soils having good or moderate ratings for wastewater introduction. Finally, while inconclusive statistically, the study did tend to indicate that systems installed in areas with higher seasonal water tables were slightly more likely to fail.

The Pelagic Food Web in Forest Lakes Affected by Alkaline Mining Waste

Industrial wastewaters often have harmful effects on recipient lakes. For example, acidic mine drainage with a low pH and high metal contents often causes serious ecological damage. Since 1984 the JSC Karelsky Okatysh mining company in the Kostomuksha area (64 [degrees]43'N, 30 [degrees]58'E) in Karelia, NW Russia, has extracted and processed magnetite ore into iron pellets (7.6 million tonnes in 2004). The company affects the environment by air pollution (SOZ and dust) and alkaline wastewater emissions. Since 1994 wastewaters (approx. 10-20 million [m^3] [y^{-1}]) from a dammed basin and from open mining pits (2-3 million [m^3] [y^{-1}]) have been allowed to flow a distance of 75 km through a number of small lakes and the Kenti River to the larger lake Middle Kuito. Waste flow represents approximately 8% of the mean discharge of 8.21 [m^3] [s^{-1}] at the river outlet. Data on water quality and biota in this water course have been given e.g. in Virtanen & Markkanen (2000), Lozovik et al. (2001), and Kalinkina et al. (2003). Preliminary results on the food web are given e.g. in Holopainen et al. (2003b).

This chapter analyses the structure of the planktonic food web in two lakes impacted by wastewaters in comparison to a reference lake. The fish faunas are similar but the catch-per-unit-effort is two-fold higher in the impacted lakes, while the water chemistry and plankton communities are drastically changed, suggesting that food chain characteristics underpin this apparently successful fish life.

Lakes Poppalijarvi (1.7 [km2], ca. 4 km downstream from the waste basin) and Kento (27.1 [km2], ca. 20 km downstream) are oligotrophic and mesohumic (Secchi depth 2-3 m) forest lakes impacted only by the mining waters. A partly isolated basin of the larger Lake Upper Kuito, flowing like the impacted lakes into Lake Middle Kuito, was used as a reference area; it is only slightly affected by the local village, Vuokkiniemi, which has approximately 500 inhabitants. The climate is continental with relatively warm summers but cold winters: all the lakes are ice- and snow-covered for some 200 days each year.

All the three lakes were sampled once in early August in three successive years (2000-2002) for water chemistry, phytoplankton, zoobenthos, and fish. In 2001-2002, bacteria, zooplankton, and Ciliata were also sampled. Samples for water quality were taken with a Ruttner-type tube sampler and analysed according to Finnish standard methods (SFS Standards) in the experienced water laboratory of the Karelian Institute at the University of Joensuu.

For bacteria, phytoplankton, and Ciliata, a sample of approximately 10 L of surface water (0-1 m) was subsampled (100 mL) and fixed with acidic Lugol solution. The hypolimnion (9-12 m, 1 m from bottom) was also sampled in 2002. Species composition and biomass (volumes) of phytoplankton and Ciliata were determined by inverted microscopy using the sedimentation chamber technique. The conversion factor from cell volumes to carbon was 1 [micro]g C [L-1] = 0.11 M [micro][m-3] [L-1] biovolume for phytoplankton and 190 fg C [micro][m-3] for Ciliata. Before the bacteria were stained with 0.20 [micro]m filtered acriflavine solution, the samples were clarified with thiosulphate crystals. Two replicate 1 [micro]L samples were filtered on black Poretics polycarbonate filters (pore size 0.22 [micro]m) and the bacteria were counted with an epifluorescence microscope at 1500 x magnification from 5-10 random fields (5901 [micro][m2]) in each sample. A video camera and the analysis 3.0 Image analysing program (Soft Imaging System GmbH, Munster) were used to determine bacterial densities and the sizes and volumes of the bacteria counted. The number of bacteria counted ranged between 636 and 836 per sample. The conversion factor from bacterial cell volumes to carbon was 360 fg C [micro][m-3].

For transmission electron microscopy (TEM), 100 mL of prefiltered (10 [micro]m) water was filtered through 0.45 pm membrane filters, which were fixed in glutaraldehyde. Two-millilitre samples of the material rinsed from the filters were then centrifuged for 5 min at 3000 xg. Planktonic material at the bottom of the tubes was fixed for 24 h in 1.5 mL 2% glutaraldehyde in 0.1 M sodium cacodylate buffer. The tubes were centrifuged again for 5 min at 3000 xg and postfixed for 3 h in 1% Os[O4] in 0.1 M sodium cacodylate buffer. After dehydration, the plankton samples were embedded in Epon. For transmission electron microscopy (TEM), ultrathin sections were cut with a diamond knife, stained with uranyl acetate and lead citrate, and examined with a Zeiss 900 electron microscope.

Zooplankton was sampled by a tube-sampler (Limnos, 6.8 L) from both littoral (0.5-2.5 m) and pelagic (0-5 m and 5-10 m separately) zones. At each of the three depths, ten lifts (68 L) were combined to form a sample, which was sieved through a plankton net (48 gm) and fixed with ethanol in the field and formaldehyde in the laboratory. The invertebrates were counted with an inverted microscope after subsampling. From a stirred sample of 250 mL, at least three quantitative sub-samples (total 60-170 mL) were pipetted into a counting chamber. The counts were converted to biomass using average carbon values for appropriate size classes of each taxon.

A stratified random sampling of fish from three habitats (bottom inshore, bottom offshore, and surface) in each lake was carried out using nine multimesh gillnets (mesh sizes 10, 12, 15, 20, 25, 30, 35, 45, and 55 mm from knot to knot in each 1.5 x 30 m net; each mesh panel 1.5 x 3 m).

All three 8-13 m deep basins sampled were unstratified in early August 2000 and 2001. In August 2002 the two impacted (smaller) lakes were stratified. Similar densities (1.8-2.8 x [106] cells m[L-l]) of bacteria were found in the surface waters of all the lakes. In the hypolimnion (9-12 m) both the densities and cell volumes were similar to those in the surface waters of Lake Poppalijarvi, whereas in the two other lakes both the size and density were lower close to the bottom. A large rod-shaped cyanobacterium (tentatively identified as Synechococcus spp.) predominated in the two impacted lakes. The TEM micrographs of this cell (approx. 1.0 [micro][m-3] or 0.7 x 2.7 [micro]m) show e.g. a four-layered cell wall. The bacteria in the impacted lakes were much larger than those in the reference lake, and because their densities were approximately the same, their biomass was also much higher.

In a routine count, the total biomass of nano-phytoplankton ranged between 16 and 78 mg C [m-3] in all lakes and years, the values being higher in the reference lake in 2000 and 2001. Cyanophyceae, Cryptophyceae, and Chlorophyceae were characteristic of Lake Poppalijarvi, and Chrysophyceae and Cryptophyceae of Lake Kento, but Diatomophyceae and Dinophyceae of the reference lake. In Lake Poppalijarvi, Cryptomonas spp. and Synedra spp. were predominant in the biomass. In 2000, however, the biomass of blue-greens (mainly Planktothrix agardhii (Gomont) Anagnostidis and Komarek, 1988) was high. In Lake Kento, small Cryptophyceae flagellates together with the chrysophytes Pseudopedinella spp. and Uroglena sp. were abundant. In the reference lake the total number of taxa was higher, and the most abundant were the blue-green Merismopedia warmingiana Lagerheim, the Cryptophyceae flagellates, and the diatom Rhizosolenia longiseta Zacharias. In 2001 and 2002, heterotrophic and potentially mixotrophic algae (Cryptophyceae, Dinophyceae, Chrysophyceae, Euglenophyceae, Choano- and Zooflagellata, unidentified heterotrophic flagellates) attained the lowest relative biomass (42-43% of total phytoplankton) in the reference lake Upper Kuito, where their absolute biomass was 34-95% of those in the impacted lakes.

In 2001 the ciliate community in the surface water of Lake Poppalijarvi was dominated by Oligotrichida (Pelagostrombidium sp. formed 50% of the biomass), but in 2002 by Peritrichia (Vorticella sp.) and Scuticociliatida. All the groups present consisted mainly of small species. In Lake Kento Oligotrichida (Rimostrombidium spp. and Limnostrombidium sp.), Prostomatida (Coleps sp.), and Hypotrichia were predominant. The high ciliate biomass in the surface water of the reference lake, Upper Kuito, was clearly dominated by Rimostrombidium spp. and Limnostrombidium sp. In the hypolimnion of Lake Poppalijarvi, 54% of the biomass consisted of Scuticociliatida; in Lake Kento Oligotrichida and Prostomatida were again predominant, and in Lake Upper Kuito Rimostrombidium spp. and Mesodinium sp. (Gymnostomatea) dominated.

The taxa richness of Rotatoria per sample increased downstream from Lake Poppalijarvi (6-10 taxa) to Lake Upper Kuito (9-15 taxa). Similarly, the numbers of Cladocera and Copepoda taxa were higher (6-8 and 4-5 taxa, respectively) in the reference lake than in the other two lakes (2-7 and 1-4, respectively). No differences were found among habitats (shore, surface, bottom) in any lake.

The total biomass of zooplankton in all areas was lowest in the most impacted Lake Poppalijarvi. Rotatoria always formed a major part of the biomass (31-80%). The dominant species were Brachionus angularis and Keratella cochlearis in Lake Poppalijarvi and the genus Keratella in Lake Kento. The pelagic zone of Lake Upper Kuito was dominated by Polyarthra sp. and Kellicottia longispina Kellicott, 1870, and the littoral by Synchaeta spp. The biomass of Cladocera increased downstream and was dominated by Bosmina longirostris in lakes Poppalijarvi and Kento, but by Daphnia cristata Sars, 1862 in Lake Upper Kuito. Despite their much lower numbers in Lake Poppalijarvi, B. longirostris appeared to be clearly (t-test $P < 0.001$) larger (length 329 [+ or -] 77 [micro]m, mean [+ or -] S.D., n = 50) than those in Lake Kento (239 [+ or -] 37 [micro]m) and to include more egg-bearing females (28/50 vs. 14/50).

The small cyclopoids Mesocyclops leuckarti Claus, 1857 and Thermocyclops oithonoides were characteristic of all three lakes. Calanoida (mainly Eudiaptomus sp.) were found regularly only in the reference lake Upper Kuito, but even there their biomass was low (max 13% of total). Both total fish catch and the catch of potential zooplanktivores were much higher in the impacted lakes.

Our singular or 'snapshot' type sampling from a single station gives limited information about the lake food web because it neglects temporal and spatial variations. Mehner et al. (2005) found the horizontal and daily variation to be less important than vertical and month-to-month variation. We sampled both the epi-and hypolimnion, and the fact that we have data from August in two or three successive years increases the reliability of the results.

The changed water chemistry in the impacted lakes is due to wastewaters from iron mining and ore processing, and the high nitrogen levels derive from the explosives used in the mining pits. The unnaturally high absolute and distorted relative ion concentrations observed since 1994 are expected to affect organisms at different levels because many cations interact with biological membranes and regulate their permeability and ion transport. Although the heavy metal contents of the wastewaters are elevated, their concentrations in the lakes downstream from the mine and their bioavailability for biota are low at the ambient high pH and water hardness.

In the impacted lakes, the biomass of pelagic bacteria and autotrophic picocyanobacteria was 2-3 times that of routine phytoplankton. The structure and function of autotrophic picoplankton (APP) communities have been reviewed by Callieri & Stockner (2002). APP appear to form a basis for microbial food webs with important effects on higher trophic levels, including fish, and may account for 10-90% of the total phytoplankton biomass or production, the highest values often being found in oligotrophic lakes. The APP biomass values in six Quebec lakes varied between 16 and 80 [micro]g C [L-1]. In a natural humic lake in southern Finland the bacterial biomass in the surface water was 14-19 [micro]g C [L-1], which agrees well with the values for our reference lake. Because of the large picocyanobacteria, the average cell volume of all bacteria in the impacted lakes was 0.09 [micro][m-3], much larger than that in the reference lake (0.01 [micro][m^3]) or that given for a humic forest lake in Finland (0.02 [micro][m-3] cell-) by Tulonen et al. (2000). In Lake Maggiore the mean cell volumes of the most abundant species of picocyanobacteria fall between 0.2 and 0.8 [micro][m-3] in August.

Chekryzheva (1995) gives a list of phytoplankton species and total biomass in 1987 and 1994 for lakes Poppalijarvi and Kento. Diatoms predominated (30-50% of taxa richness) together with Chlorophyceae (20-30%) and Chrysophyceae (15-25%). In 1994 the biomass was higher in Kento than in Poppalijarvi (325 vs. 150 [micro]g [L-1]), whereas our results suggest similar biomasses for these two lakes.

In 1994 the number of taxa was clearly higher in Lake Kento than in Lake Poppalijarvi. According to Arvola et al. (1999), the average number of phytoplankton taxa per sample in oligotrophic Finnish lakes is 54-67. The relative proportions of blue-green and green algae were highest in Lake Poppahjarvi. Chrysophyceae are negatively affected by high potassium content: Dinobryon spp. stop growing in water at 15 mg [K^+] [L^{-1}], but high concentrations of the divalent cations [Ca^{2+}] and [Mg^{2+}] increase their tolerance to high [K^+]. In Lake Poppahjarvi Chrysophyceae were absent from all samples, except for a small Pedinella species. In Lake Kento Chrysophyceae predominated, but their species richness was low (7-8 taxa) compared to that in Lake Upper Kuito

(10-16 taxa). In 1994 Chrysophyceae were well represented in all lakes, even upstream from Lake Poppalijarvi. According to Kalinkina et al. (2003), potassium increased in Lake Poppalijarvi from 10.6 mg [L^{-1}] in 1993 to 42.4 mg [L^{-1}] in 1994 and to 60 mg [L^{-1}] in 1995, accompanied by some increase in [Ca^{2+}] and [Mg^{2+}].

The chlorophyll a values (2-4 [micro]g L-) in the reference lake are compatible with those in other oligotrophic lakes in the region. In 1990-2001 the July-August wet biomass of phytoplankton in single annual samples in Lake Iso Hietajarvi varied between 0.1 and 0.4 mg [L-1]. Our biomass values for all three lakes conform closely to these results. However, in the impacted lakes, Poppahjarvi and Kento, the chlorophyll a content and the biomass of primary producers were much higher because of the high densities of picocyanobacteria, and owing to their small size and fast turnover rate the effect of these organisms on production could be even greater.

In 2001 the total ciliate abundance in the surface water of Lake Poppalijarvi was higher than in some oligotrophic Finnish lakes, but still in the range typical of an oligotrophic environment. The predominant Pelagostrombidium species feeds on nanoplankton. In 2002 the bacterivorous Scuticociliatida were dominant in both the epilimnion and hypolimnion. The other groups present were also dominated mainly by small species that are known to feed on picoplankton. This suggests that microbial processes are important in Lake Poppalijarvi, especially in the hypolimnion, where the highest abundances were found. The picoplanktivorous nanoflagellates and Ciliata liberate the nutrients bound in the picoplankton biomass by grazing, thus mediating their availability to higher trophic levels.

Lake Kento had higher ciliate abundances in the epilimnion than Lake Poppalijarvi, but much lower numbers in the hypolimnion. The predominant species (Rimostrombidium spp.) is a nanoplankton feeder. The abundances as well as the community composition were typical of oligo and mesotrophic lakes. Both in 2001 and 2002 the ciliate community in Lake Upper Kuito was similar to that in Lake Kento, and in 2002 the differences between the epi- and hypolimnion were small.

The ciliate species richness in the reference lake was similar to that in oligotrophic Finnish lakes, but was lower in the impacted lakes, especially in Lake Poppalijarvi. Nevertheless, the community structure in Lake Poppalijarvi resembled that found in eutrophic systems, which are usually rich in ciliate species.

The results obtained by Kalinkina et al. (2003) agree well with those given in Vlasova (1998) and show biomass values between 80 and 800 mg ww [m^{-3}] for zooplankton in Lake Poppalijarvi in July-August 1992 to 1997, but only 12-15 mg ww [m^{-3}] in 1999 to 2001. Our results for the pelagic surface water

(0-5 m) show that the zooplankton biomasses are highest in Lake Kento and Lake Upper Kuito (up to 850 mg ww [m^{-3}]; fresh mass is carbon mass x 18.093). The biomass of Lake Poppalijarvi was the lowest (224 and 172 mg ww [m-3] in 2001 and 2002, respectively) but still an order of magnitude greater than those given by Kalinkina et al. (2003) for 2001. This could be due, at least in part, to the large mesh size (99 gm) and consequent lack of small Rotatoria in their results. Vlasova (1998) stated that Rotatoria were counted separately from unfiltered samples; there is, however, no mention of this in Kalinkina et al. (2003).

The average biomass of crustacean zooplankton in 138 South Finnish lakes was 40 mg C [m. Owing to the predominance of Rotatoria and the low numbers of Calanoida (and Cladocera in L. Poppalijarvi), the carbon mass of Crustacea was much lower in our lakes, with the highest values found at a depth of 5-10 m. According to Kalinkina et al. (2003), Calanoida (mainly Eudiaptomus gracilis Sars) and other species favouring soft waters practically disappeared from Lake Poppalijarvi in 1995. This was shown experimentally to be due to intolerance to the high (> 50 mg [L.sup-1]) potassium content. In our results the abundance of Calanoida was exceptionally low, even in the reference lake. In 1994 the zooplankton of Lake Poppalijarvi was dominated by Daphnia cristata and D. longispina, but in 1996 by Bosmina longirostris. No marked changes have been detected in zooplankton of the natural Lake Middle Kuito.

Some trophic relationships that have commonly been used to characterise food chains in lakes suggest top-down control of zooplankton by fish in the impacted lakes. However, this material is not complete enough (e.g. lack of data on heterotrophic nanoflagellates) to explain all details of the microbial food web in the pelagic zone, or the possible functional coupling between the abundant picocyanobacteria and the large fish biomass. Furthermore, the predominant fish (roach) is benthi-planktivorous, and the higher biomass of benthic animals in both littoral and profundal zones of the impacted lakes may be more important than the pelagic zooplankton for fish density. Recent studies have suggested that fish are strongly dependent on the littoral zone and there might be only a weak connection between the littoral and the pelagic food webs in oligotrophic lakes.

8

Hydropolitics: Water and Security

The semiarid environment of the Indus Basin is home to more than a quarter of a billion people with some of the lowest human development indicators in the world. As if the marginal environment and the pervasive poverty were not enough, deep political fissures across international, subnational, and local boundaries characterise the political geography of the basin. The basin would be a hostile desert if it were not for the Indus Basin Rivers and the largest contiguous surface irrigation system in the world emanating from those rivers. Needless to say, just as Egypt has been described as a gift of the Nile, the bustling ancient cultures of northwestern South Asia and present-day Pakistan and northwestern India can be called a gift of the Indus. Given the stakes involved, in terms of the survival of millions of people, the Indus River Basin has been a veritable laboratory for international and national research on various problems with water distribution, development, and management, especially as they pertain to issues of water efficiency, equity, hazards, and environmental quality.

The story of water resources in the Indus Basin is intricately linked to the political geography of South Asia, particularly in the colonial and precolonial times. But much of the attention to the hydropolitics of the Indus Basin has been either through a historical lens or limited to the international scale, and very little research has been conducted on the contemporary subnational levels of prevalent and potential water conflict in the basin. Is there a nexus between security and hydropolitics? What are the implications of a security-centred approach to hydropolitics? What are the implications of conflict over water across local, interprovincial, and international scales? How does the geography of water-resources distribution, development, and hazards at the subnational scale contribute to or threaten security at the national and international scales? This chapter defines a research agenda in the Indus Basin, in the hope that it will provide a road map for future research to address some of these questions.

The Water-Security Nexus

The end of cold war ushered in an era of intellectual instability in the field of security studies. Whereas during the cold war the relative certainties of a competitive, bipolar world kept the issue of security the exclusive preserve of the military and foreign-policy establishments, in the post-Soviet world a variety of new agendas have been subsumed within the security discourse, including the environment, economy, and, more recently, terrorism. The new concern with environmental security is premised on a wider understanding of the concept of security, one that goes beyond the traditional realist and neorealist theories of security that privileged power relations, especially military power and, to a lesser extent, economic power over all other aspects of national and international security. Those scholars and activists who are concerned with wider conceptualisations of national and international security argue that environmental degradation and resource depletion can threaten social, economic, and political stability and may very well lead to civil and military conflict.

At the international scale considerable attention has been paid to the probability of water wars in the future. Aaron Wolf argued that, based on empirical evidence, transboundary water conflicts are much more likely to be resolved through collaboration than through armed conflict (1997, 2002). Other scholars, such as Hussein Amery, who analyses the short-lived hysteria in Israel over a local water-development project in Lebanon on the Wazzani Spring, a tributary of the Jordan River from which Israel draws 60 percent of its water resources (2002), argue that conflict over water continues to have the potential for turning violent. More recently, Mark Zeitoun and Jeroen Warner have drawn attention to the fact that lack of overt armed conflict does not mean that asymmetries of power at the international level do not play a role in water management (2006). Drawing primarily on case studies of the Jordan, Nile, and Tigris and Euphrates River Basins, Zeitoun and Warner argue for a hydrohegemony framework in which the most powerful riparian imposes unfavourable water agreements on weaker riparians by threatening to use force or through superior bargaining or discursive power.

This chapter is premised on the belief that security does indeed involve more than just military or economic aspects, and its implications are not just in terms of threats to the political order and organisation of nation-states but also involve very real threats to the life, property, and peace of human populations. Environment and resources, being factors in providing the material basis for human existence and social life, are tied up with issues of security insofar as lack of them and conflict over them threatens the well-being of human populations.

Thomas Homer-Dixon is one of the more influential proponents of the environment-security nexus, but his formulation of the nexus describes population growth as an integral-causative component of the nexus, where environmentally induced conflict and violence are mediated by environmental scarcity. The population, environmental scarcity, and conflict model has become the dominant paradigm for understanding the environment-security nexus. In his later work Homer-Dixon posits that maldistribution of resources, environmental degradation, and population growth can equally contribute to potential social instability and conflict (1999). The demographic-pressure part of the model is problematic at best and counterproductive at worst. Scholars like Betsy Hartmann point to analytical obfuscations, devaluation of civil society, legitimating and normalising of injustice, and even sanctioning of thinly disguised racism and sexism as the main pitfalls of accepting the model (1999). Many writers have also outlined detailed comments against accepting population growth as the driving force for environmental degradation. Even in Homer-Dixon's formulation, the weight of evidence points to distributional inequities in causing "ecological marginalisation" of the majority of the unprivileged in most societies, yet, somehow, absolute demographic pressures and average resource distributions per capita continue to feature prominently in his conclusions; for example, "Ecological marginalisation occurs when unequal resource access combines with population growth to cause long-term migrations of people dependent on renewable resources for their livelihood".

If the bulk of environmental conflict and instability is indeed the outcome of distributional factors, what is the point of smuggling the old wine of population growth as a driver of resource depletion in the new bottle of "environmental scarcity"? To many scholars—for example, Richard Peet and Michael Watts (1996)—to switch focus from the political economic factors that affect access to resources is, in fact, tantamount to turning a blind eye to the injustices at the heart of producing affluence for the few at the expense of scarcity and misery for the many. Nowhere is the need for a focus on the political economic and discursive factors driving resource use and distribution more urgent than in the field of water resources. The sterile per capita freshwater availability numbers may seem alarming—as they do to Peter Gleick (2000), for instance—but they really serve to divert attention from water's problematic social geography, from its extremely skewed distribution across sectors and across social groups, and from discursive construction by the power elites as a "resource" to be deployed in isolation from its ecological and social roles toward modernist economic development. It is surprising that ordinary water users at the local scale tend to know that water scarcity is really mediated by social power relations, yet scholars continue to talk about per capita numbers.

The following analyses of water and security in the Indus Basin reject the centrality of the neo-Malthusian population-growth-based explanations for resource scarcity and environmental degradation and conflict. Instead, the chapter draws its intellectual capital from the twin theoretical positions that environmental degradation, resource scarcity and security are all socially constructed—normative and collective understandings that have consequences for physical and social worlds. Furthermore, the epistemic—knowledge-based experts—and political communities that are most influential in the social construction of environment and security are to be found at the subnational level, but with important linkages to international epistemic communities (for example, the engineering profession). The chapter's theoretical position is therefore consistent with the call by Christopher Sneddon and Coleen Cox for a critical hydropolitics that, in the context of their case study of Mekong River Basin, helps to "reveal barriers—discursive, political and institutional—to sustainable governance and meaningful participation" (2006, 198). The following survey of the Indus Basin concentrates on the subnational-level social construction of hydropolitics and its links with subnational and international security.

Indus Basin Hydropolitics

The Indus Basin has been host to irrigated agriculture for at least five millennia, but none of the precolonial water development matched the environmentally and socially transformative power of the water development undertaken by the British colonial government in the later half of the nineteenth century. Coincidentally, much of the massive water development undertaken by the British in the Indus Basin was motivated at least partially by national security considerations, and many of the consequences and conflicts arising from the development of the Indus Basin irrigation system were viewed through the lens of the Raj's security in northwestern India. The British colonial administration had several motives for massive water development in the Indus Basin:

- to increase food production in order to prevent famine;
- to anticipate increased tax revenues from the potential increase in agricultural production, which was expected from water development;
- to increase government control of the local populations by encouraging them to take up settled agriculture and thereby minimise the security threat they might pose to the power of the state;
- to demonstrate to local people Western science's control over the environment, thereby discouraging them from posing any threat to the security of the empire; and

- to create and develop new social elites through the settlement policies that were to follow the water development—elites who, because they owed their material and political power to their connections with the British Empire, would further secure British rule.

Needless to say, even from a narrow realist perspective of military and political security the environment-security nexus was very much in the minds of nineteenth-century water developers and managers. David Gilmartin documented at some length the importance of local-level water management to the patron-client relationship that the British Empire had developed with the local elites and the importance they attached to that relationship, as a guarantor of the security of British rule in northwestern India (1994). In fact, the nineteenth-century history of water development and management in the Indus Basin is a story of constant conflict between the security-minded civil administration, which favoured the privilege of the local elite, and the water engineers who wanted science and engineering to be the fundamental criteria for water management. De facto, the security-and-stable-governance-driven agenda of the civil administration generally prevailed over the technocratic agenda of the engineers, both in the colonial period and in the postcolonial period.

In addition to concern with local level implications of security, Indus Basin water development, particularly in the upper portion of the basin, engendered conflict at the regional scale. Very early in the history of water development in the upper Indus Basin the downstream province of Sindh, which was at the time part of the Bombay Presidency, started vigorously objecting to further water-development projects in the upper basin. Although the conflict was generally limited to bureaucratic wrangling between the water bureaucracies of the two provinces directed toward undermining each others' water projects, it was a harbinger of what was to come in the postcolonial Indus Basin. However, the British government was quite sensitive to the implications of the conflict in the atmosphere of nationalist struggle in South Asia during the late 1930s and 1940s. The themes of interprovincial and local-level conflict over water and its implications for national and international security were to carry over into the postcolonial period and were to define the water-security nexus in the postcolonial Indus Basin.

In the immediate aftermath of the partition of the subcontinent between the two independent states of Pakistan and India, the more urgent issue of water distribution between the now-divided Indus Basin eclipsed the interprovincial water conflict between the Pakistani provinces of Sindh and Punjab. Downstream Sindh and upstream Punjab redirected much of their historical hostility over water issues toward the Indian government's plans for water-development projects on the headwaters of the Indus tributary rivers running through its territory. The Indian government was also grappling with the conflict over the distribution of water among the states of Punjab, Haryana, and Rajasthan. Thanks to the active mediation and financial support of the World Bank and the Western powers led by the United States, India and Pakistan signed the Indus Water Treaty in 1960, allocating the entire flow of the three eastern tributaries of the Indus River to India and that of the three western tributaries to Pakistan. The World Bank rewarded (in a manner of speaking) both Pakistan and India with massive aid inflows to build storage and conveyance facilities to provide remedial water supplies for the flows that were supposedly lost to the other country.

The massive water development carried out in both India and Pakistan as part of the Indus Basin Water Development Project in the aftermath of the Indus Waters Treaty provided a very temporary boon to agricultural water supplies in the basin. However, with the international aspect of the water-distribution issue settled for the time being, the old issue of interprovincial water distribution slowly crept back into the forefront, both in India and in Pakistan, though much more forcefully in the latter.

Furthermore, the changing demographics of the basin, with greater urban-based populations, and the emergent widening of the international water-resources agenda from a singular focus on water development to equity in water resources, water management, environmental quality, and domestic water supply and sanitation meant that the old recipe of responding with more water-development projects in the face of water demand and hazards was no longer as readily available. The water bureaucracies in the basin, however, being dominated by engineers with a very strong institutional bias toward megaprojects, continue to operate as if the multiple-point agenda with regard to water resources can be addressed only through additional engineering interventions. Much of the hydropolitics of the Indus Basin and the conflicts arising from it can be understood, at least partially, with reference to this dissonance between the multidimensional expectations and demands for water resources by the public, international donors, and politicians and the unidimensional solutions offered by the basin's water managers.

Ayub Qutub and his coauthors offer a typology of water-related conflict based on scarcity, technical capacities and human demands, political factors, and institutional/legal informational inadequacies (2004). Based on the typology of conflict and a survey of various water conflicts in South Asia, they suggest that water conflict may be "fractal" in nature; that is, having repetitive similar characteristics across geographical scales, with the most important manifestations at the local scale. The chapter recognises the multidimensional, perhaps even fractal, nature of water-related conflict but proposes that all types of water conflicts are nested within specific material and discursive social structures that may operate across spatial scales. The following analysis of water and security nexus at the subbasin level is an attempt to apprehend the contours of those structures, in order to inform public policy and research.

Tracing the Contours of Subnational Interprovincial Hydropolitics

In both India and Pakistan, interprovincial hydropolitics have been political lightening rods in terms of interprovincial relations. In the case of India the issue of interstate water distribution became one of the catalysts for a very destructive separatist insurgency. In the case of Pakistan, however, the conflict over water distribution between the dominant Punjab Province and the other, smaller provinces in the federation, particularly Sindh Province, has remained peaceful and limited to the political arena, although its wholesale appropriation by the Sindhi nationalist elements in their rhetoric bodes ill for the future. I review the Indian example in order to evaluate the prospect of Pakistan's heading down the same unfortunate path as the Indian Punjab because of the simmering controversy over the construction of additional storage on the Indus River.

The details of separatist insurgency in the Indian Punjab can be found elsewhere (Singh 2000). Suffice it to say here that it, like all civil conflicts, had a multiplicity of contributing causes, including ethnoreligious identity politics, the question of distribution of water and other resources among the states of Punjab, Haryana and Rajasthan, and the mishandling by the central government of India of the Sikh grievances, thereby turning the insurgency into a full-fledged armed conflict. Of the many causes, the issue of water distribution among the states of Punjab, Haryana, and Rajasthan came to be very liberally used by the Sikh nationalist elements in their rhetoric against the central government of India. To briefly recap the history of the conflict, in 1966 the Indian government carved Haryana State out of the southeastern portion of Punjab State along linguistic and religious lines.

In 1955 an interstate agreement dividing the waters of the Sutluj, Ravi, and Beas rivers among the still-undivided Punjab State, Jammu and Kashmir State, and Rajasthan State had ushered in an era of extensive water development on the Beas and Sutluj rivers. The victory of the Sikh nationalist Akali Dal Party in the Punjab in 1967 further compounded the conflict between the states of Punjab and Haryana. Punjab insisted on exclusive control of the Beas and Sutluj waters because the two rivers were exclusively within its territory, while Haryana demanded an apportionment of the waters based on "needs and principle of equity". Liberal use of the water issue to inflame public opinion in the Punjab, coupled with the central government's power politics, widened the schism between the Sikhs and the national mainstream in India, resulting in the tragic loss of lives in a brutal civil insurgency throughout the 1980s.

The water conflict among Punjab, Haryana, and Rajasthan states can be viewed in the context of single-minded commitment on part of the Indian government to massive engineering interventions for water development. The Pong Dam on the Beas River, the Indira Gandhi Canal from the Harike

Barrage on the Ravi, the Beas-Sutluj Link Canal, and the Bhakra-Nangal Dam project on the Sutluj, to name a few of the gigantic water projects, have completely rendered the hydrology of the Indian Punjab, much like its western counterpart in Pakistan, more cultural and political than natural. The motivation of the Indian water managers, much like that of their Pakistani counterparts, was to maximise the development of the water resources and put them to narrowly (economically) defined beneficial use.

Many questions present themselves when trying to analyse the role of water in instigating a fratricidal conflict that almost spun into an international conflict between India and Pakistan when Indian armed forces were mobilised in 1987 on the pretext of stopping Pakistan's alleged support of militancy in the Indian Punjab. Did a dissonance exist between the dominant technocratic view of water with a single-minded focus on large water-development projects and the Sikh society's wider cultural, spiritual, economic, and social values placed on water? What role did the dissonance play in further fanning Sikh militancy in the Punjab? What was the role of the Punjab rivers in the identity politics of the Sikh population in the Punjab? And could it be that the interstate water-distribution issue came to be linked to the politics of ethnoreligious identity? What role did the green-revolution technologies play in creating the massive demand and therefore conflict over water? To what extent was the water conflict underpinned by concerns for equity and food security as opposed to surplus accumulation from cash-crop production? These are some of the questions that future research needs to address and that I revisit in the context of the Kalabagh Dam controversy and the interprovincial water conflict over the Indus river waters in Pakistan. The former is a proposed dam on the Indus River near the town of Kalabagh in northwestern Punjab. The project has been on the drawing board since the 1960s, but vigorous opposition from the three less-populous provinces of Pakistan has kept the project in abeyance.

As I mentioned above, in Pakistan the interprovincial conflict over the allocation of the Indus River's waters dates back to the beginning of massive canal construction by the British in the Punjab in the mid-nineteenth century. The first substantial interprovincial water-allocation treaty between the Punjab and the downstream riparian Sindh provinces was signed in 1945. The treaty allocated 75 percent of the waters of the main-stem Indus River to Sindh, with the remainder going to Punjab. The treaty further allocated 94 percent of the water from the five eastern tributaries of the Indus River to Punjab, with the residual water going to Sindh. The partition of the subcontinent and the subsequent signing of the Indus Water Treaty by India and Pakistan in 1960 allocated most of what was Punjab's share of the Indus Basin waters according to the 1945 Sindh-Punjab Agreement to India and provided for construction of storage and link canals from the western half of the Indus Basin to the eastern half to compensate for the water lost to India. The Sindhis widely perceived compensatory water and the storage on the Indus and Jehlum rivers to be compensation to Punjab at the expense of Sindh (Talpur 2001). The Kalabagh Dam controversy is proving to be yet another insult to the long series of injuries that Sindhis perceive to have been inflicted on them by Punjabis because of their appropriation of Sindh's rightful share of water.

The seemingly perpetual water conflict between Sindh and Punjab had a tentative settlement in the form of the interprovincial water accord of 1991, when the four provincial governments, all of which were governed by the same political party, for the first time agreed to a water-allocation formula. Based on the assumed average flow of 114.35 million acre-feet (MAF) of water in the Indus system, the accord allocated 55.94 MAF of water to Punjab Province and 48.76 MAF to Sindh province (PWG 2003). Although Humaira Afzal argued that the actual apportionment came closest to what a reasonable apportionment could be, the accord nevertheless suffered a crisis of legitimacy (1995). Legitimacy was in question primarily because the negotiating process leading up to the accord was not transparent and did not include all the stakeholders, particularly from the smaller provinces, and secondarily because of the suspect legitimacy of the political setup in Sindh at the time. Furthermore, even the official statistics for average annual flows for the Indus Basin used in the interprovincial water accord and subsequent justifications for additional storage on the Indus River, particularly the Kalabagh Dam, are suspect. Many writers, including Shaheen Khan (2003) and Abdul Kazi and A. N. G. Abbasi (2003), convincingly argue against the official methodology of using the higher number for flows in the Indus system, particularly because it is based on a shorter time frame—that is, since 1977—and because the higher number works to the disadvantage of downstream, riparian Sindh Province. The official argument in favour of the construction of the Kalabagh Dam on the Indus River paints the picture of a scarce water resource that is being wasted by being allowed to flow out to the sea and outlines a doomsday scenario should additional storage not be built (Government of Pakistan 2005). The controversy is beginning to polarise public opinion in Pakistan, particularly in Sindh Province.

On the internal-security front, water scarcity in Sindh Province, especially in the aftermath of the drought experienced in southern Pakistan in the later half of the 1990s, coupled with the single-minded focus of the Pakistani water bureaucracy on water development, has made the Kalabagh Dam project a surrogate for a litany of Sindhi grievances against the Punjabi-dominated political, military, and bureaucratic system in Pakistan. On the other hand, for the Pakistani water managers, the Kalabagh Dam has become a metaphor for the persistent meddling of "untrained" and "nonexpert" politicians in what they perceive or wish to be a purely engineering issue. All types of appeals to patriotism, science, economics, and neo-Malthusian scenarios are being pressed into service by the Pakistani government and the engineering establishment to make the case for the Kalabagh Dam, as well as for other storage projects on the Indus. The dam project at the moment is in cold storage, particularly because of the combined opposition of not just Sindh Province but also of the North Western Frontier Province (NWFP) and Balochistan Province. The NWFP is concerned lest the lake that will be created behind the dam flood rich farmland and the Pashtun cultural heartland. The province is also reluctant to lend its support to the project because of suspicions based on the poor record of the Pakistani government in providing for the rehabilitation of dam affectees from earlier large dam projects.

The objections to the additional storage issue on the Indus River are not limited to the nationalist politics of smaller provinces. Other very convincing arguments have been made by environmental and citizens groups in Pakistan, pointing out that Pakistan's irrigation sector has some of the lowest conveyance efficiencies in the world. The detractors argue that, instead of going for very expensive, environmentally damaging, and economically dubious storage and megaproject solutions to the water issue in Pakistan, perhaps enhancement of the existing infrastructure's efficiency, coupled with better on farm water management and more appropriate irrigation and farming techniques, would result in more water than would be gained from megaprojects. Clearly, additional rigorous research on the subject will be useful in informing water-policy debates in Pakistan.

At the moment, the Indus Basin water-distribution controversy is limited to sloganeering and street protests on part of the populace of Sindh and, to a lesser extent, the NWFP, as well as to heated debates among the water managers and provincial governments of the Pakistani federation. Incidentally, in Punjab, the province that stands to benefit the most from construction of the Kalabagh Dam—and other water-development projects, such as the Greater Thai Canal project, which is to supply additional water from the Indus to the Thai area of Punjab—public opinion at the grassroots level is disinterested at best, unlike that in Sindh. This is one controversy in which engineers' concepts of how to manage and develop water, rather than any popular Punjabi demand for additional water projects, seems to be driving the conflict. For example, the currently ruling, Punjabi-dominated Pakistan Muslim League (PML-QA) has gone to great pains to try to mobilise grassroots support for the dam—with little evidence of success. This is in stark contrast to the Indian situation, in which public opinion was inflamed in support of keeping the waters of Punjab State waters from Haryana State. Whereas in Sindh hydropolitics may have fused with identity politics of Sindhi nationalists, in the Pakistani Punjab popular passions regarding the hydropolitics do not seem to exist. The shallowness of popular support for additional water development on the Indus River provides an opportunity for a more enlightened and multidimensional policy dialogue to resolve the controversy. Greater action research on some of the questions listed in case of the Indian Punjab will serve to inform the water-policy dialogue in Pakistan and will provide the basis for conflict resolution. The specter of an Indian Punjab-style insurgency with hydropolitics as one of the key issues is a nightmarish, but avoidable, scenario for Pakistan if the parameters of the discourse are widened from purely engineering concerns to wider, social, cultural, environmental, equity, and justice-related concerns over water resources.

Hydropolitics at the local scale in Pakistan are also driven by the dissonance between water users' need for access to affordable and safe domestic water supply and to their legally sanctioned irrigation water rights to fulfill livelihood needs, and water managers' focus on megaprojects for water supply and engineering-based optimisation of the irrigation system. I briefly discuss the two sectors, the former with reference to water-supply issues in Karachi and the latter with reference to the role of social power in influencing vulnerability to flood hazard and access to irrigation water at the local scale in Punjab Province.

Karachi has been described as Asia's roughest town: Between 1993 and 1996, 4,091 people sustained violent injuries in the city, of which 2,400 were fatal, and, after a tenfold decline in homicide rate in 1999/2000, the homicide rate in Karachi stood at 555 in 2002. Homer-Dixon presented Karachi as a prime example of how conflict over a scarce resource like water in conjunction with other factors—population growth and ethnic differences, for example—may lead to the type of very violent ethnic and religious conflict that Karachi experienced during the 1990s (1999). I contend that it will be more useful to view the ethnic conflict in Karachi as a failure of governance rather than as the outcome of an inexorable synergistic relationship between population growth and resource scarcity. The Malthusian explanation may be attractive because of its simplicity, but I am afraid that it may detract from more fundamental social structural causes. Qutub documents the role of the water-tanker mafia in Karachi and its role in creating an artificial scarcity of drinking water in the interest of private profit (2005). Some of the complexities of the social and environmental milieu in Karachi in the context of globalisation that he documents further caution us against viewing Karachi's problems through a monochromatic Malthusian lens. Moreover, almost every major city in the Third World is suffering from high population growth rates and issues of competing ethnic identities and lack of civic amenities, but not all cities explode into the type of violence that Karachi witnessed in the 1990s.

Domestic water supply is one of the key responsibilities of any city or state government, and the government of Pakistan and the municipal government of Karachi have consistently failed in that responsibility, especially in the poorer neighbourhoods of the city. In the context of Karachi, then, lack of access to clean drinking water is indicative of the failure of governance on many other fronts: transportation, representative governance, sanitation, infrastructure, health (Marsh and others 1995), and other sectors of civic life that make for a livable city. The coupling of each of the spheres of civic life with the question of ethnic identity both discursively and numerically—for example, the monopoly of one ethnic group over the transportation sector and the monopoly of another over law-enforcement agencies—makes for a volatile sociopolitical environment indeed. Furthermore, the fact that, at times, the price paid for water in the poorest neighbourhoods of Karachi is up to sixty times the price paid by residents of the more affluent neighbourhoods, which have a regular water connection, is a poignant symbol of social injustice. While most of Karachi's poorer localities continue to be underserved, a new desalination plant is to come on line for its posh Defence neighbourhood, and plans are under way for three more desalination plants to be built for the city. Indeed, not only have the city and provincial governments consistently underinvested in larger water infrastructure projects, with the exception of the desalination plants (to the extent that the water supply to the city has been consistently up to 25 percent below the average demand since 1990), but that underinvestment has also been exacerbated by endemic corruption, operations and management problems, and deliberate breaking of the pipes, mostly by shantytown dwellers, in order to gain access to water, or by influential elements.

More qualitative and theoretically driven research on the urban geography of Karachi, with governance and hydropolitics as the key foci, may help elucidate the social context within which hydropolitics are constructed and experienced, with potential consequences for security. In the irrigation sector, my local-level research documents the role of social power in influencing access to irrigation water and in vulnerability to flood hazard in the Khanewal District of Punjab Province. The case study illustrates how, even at the local level, water scarcity and vulnerability to flood hazard are mediated through social power relations, backed up by the coupling of the state power and extremely skewed property-ownership patterns. In a gravity-based irrigation system like that in Pakistan, with supply-based, fixed-time rotational irrigation scheduling, location is a key component in determining access to irrigation water. The research demonstrates a correlation between location, socioeconomic status, and social power, with the result that farmers who are locationally disadvantaged also happen to receive less than their share of water and are also more exposed to flood hazard. The consequences of this long-term, institutionally determined water scarcity mean that more and more smaller farmers in Khanewal District are finding it difficult to continue as agriculturists and are increasingly migrating to the shantytowns in the urban areas of the rural districts and beyond. One of the most deleterious consequences of this trend is the rise in religious fanaticism, partially fanned by powerful vested interests such as the Pakistani spy agency, which used the armies of poor, recent rural-to-urban migrants to fight the jihad in Kashmir and Afghanistan. Khanewal District is one of the major centres of sectarian organisations in Pakistan. Lack of access to water is neither the only nor the central causative element in the trend of increased religious fanaticism and rural-to-urban migration, but it is one of the contributing factors.

Much of the research on Pakistani irrigation system has stressed the importance of ensuring equity, system efficiency, and the social aspects of irrigation management. The Pakistani water-management bureaucracy, however, been more concerned with megaprojects, law-enforcement issues, and technical aspects of irrigation management rather than with social issues that impact system efficiency and inequities in access to water. As I document (2001, 2002b), not only is the colonial enabling legislation for bureaucratic management of the irrigation system patronizing and indifferent to geographies of differential access to water, the actual water managers tend to see their roles as law enforcers and technical engineers rather than as service providers. The actual water users, on the other hand, are quite mindful of the institutional distortions that influence their access to water and vulnerability to hazards.

Further research on how hydropolitics at the local level feed into local-level social trends of militancy and rising crime will go a long way toward identifying or making a case for social and water-sector reform, which may stop or even reverse the trends. Meaningful land reforms, increasing the conveyance efficiency of the irrigation system, and participatory irrigation management are some of the policy reforms for which a case could be made based on further research on the security aspects of institutionally induced water scarcity.

Toward a Social Construction of Hydropolitics Across Spatial Scales

At the subnational scale, considerably more research needs to be undertaken to establish the relative probability of violence versus cooperation as a means of conflict resolution. Furthermore, as I have argued in this chapter, water issues in the Indus Basin, much like everywhere else, are typically nested within broader issues related to democratisation, resource distribution, social justice, ethnic, religious, and linguistic identity, and economic well-being. With the complex interconnections that contribute to security—broadly conceptualised —teasing out the specific role of water in enhancing or diminishing national, human, economic, or cultural security is extremely difficult.

Obviously, given the central role of water as the basis of all life, at the intuitive level it is critical, but human institutions and discursive constructs mediate access to it and to the multiple values—economic/utilitarian, esthetic, cultural, spiritual, and so forth—that societies hope to derive from it. Just the fact that more than 95 percent of water withdrawals in the Indus Basin are dedicated to agriculture where its efficiency does not exceed 36 percent is a clear indicator that the scarcity of water is institutional rather than absolute. Increased efficiency with which irrigation water is used through engineering as well as institutional reforms, coupled with intersectoral water transfers, may more than make up for any water scarcity. Consequently, the question of whether water shortages and inequities in water distribution will lead to violence or threaten human security also becomes contingent on how water-related institutions behave.

Important lessons can be drawn from this overview of water and security in the Indus Basin. First, the dissonance between the single-point, engineering-based agenda of the water managers, which ends up contributing more to cash-crop production and further integration of local and national economies into the global economy, and the wider agenda of the water users is one of the main contributors to conflict over water, as I demonstrated in my discussion of the Kalabagh Dam. Second, the issue of access to water is intricately tied up with the issue of responsive governance, as I described in the Karachi example. Lack of good governance threatens human security by limiting access of the populace to many amenities, water being the most critical. Attention to the governance-water nexus can considerably enhance security while it improves the people's access to this critical resource. Third, hydropolitics are as much power politics as any other politics; take my work in Khanewal District. Specific attention to social power as a critical variable influencing equitable, productive, and environmentally benign uses of water is a must. Lastly, if issues of water distribution are left alone to fester, they can become a dangerous tool in the hands of political leaders who espouse other ethnic or linguistic agendas, as I showed in the case of Punjab State or as Amery documents in the case of Israel and Lebanon (2002).

The emotional and cultural appeal of the water issue is such that its misuse can cause considerable grief for those who choose to ignore it. My hope is that this chapter provides useful research directions for understanding the nexus between water and security. The overarching theme that seems to emerge for me is that enough water exists to meet the needs of populations considerably larger than those that currently inhabit the Indus Basin. However, given the vexingly uneven distribution of water across space and time, human institutions and ingenuity will be the final arbiters of whether this precious resource, the basis of physical, cultural and spiritual life, will be the basis for cooperation or for conflict in the future.

Qutub and his coauthors, in using the term "fractal" to describe water conflict at and across geographical scales, are perhaps drawing on the definition of fractals as geometric objects that are infinitely complex, have fine structures at smaller scales, have irregular shapes that cannot be described in traditional Euclidean geometry, and are recursive or self-similar (2004). Although the term "fractal" would not have been my choice of a word to describe water conflict—and it is not in this chapter—I believe that Qutub is employing a useful though obscure term to draw attention to the complexity and recursive nature of water conflict at different scales; for example, the ubiquitous notion of water as a fungible resource, common to the contemporary understanding of water at different scales, underlies most water conflicts.

In addition, no political party of any significance in Pakistan in general and in the Punjab in particular has ever passed a resolution in support of the Kalabagh Dam; nor does any evidence exist of a public rally held in support of the dam in Punjab. In the aftermath of the Pakistan's nuclear tests in 1998 the Punjab Assembly did pass a resolution congratulating the then Prime Minister Nawaz Sharif on the nuclear tests and on announcement of the construction of the Kalabagh Dam. No assembly resolution has specifically supported construction of the dam, however. The system is gravity based and therefore insensitive to water demand during various phases of crop growth and/or to the varying water requirements of different crops. The fixed-time rotational irrigation scheduling implies that every farmer along a watercourse has a claim to the entire flow of the watercourse for an amount of time proportional to his/her landholding. The farmer at the head of a watercourse may have, for example, two hours of water right, after which it would be the turn of the next farmer along the watercourse for a certain amount of time, and so on. In such a system, a water user at the head of a watercourse has a distinct advantage over a tail-end water user, for the latter may not receive any water at all when his or her turn finally comes because of seepage and evaporation losses along the watercourse.

The Global Water System

Public policy on water has long been approached in the context of a locality, a country, or a river basin. However, scientific evidence now provides compelling arguments for adopting a global perspective on water management. This chapter argues that water governance today needs a multilevel design, including a significant global dimension. The discussion defines global water governance, highlights the implications for multilevel governance, and examines global water governance through the lens of governance typologies. The analysis along the categories of globalisation/regionalisation, centralisation/decentralisation, formality/informality, and state/nonstate actors and processes reveals that current global water governance is a fragmented, mobius-web arrangement. The chapter concludes by considering possible future trajectories of global water governance. Keywords: global water governance and institutions, multilevel governance, global environmental change. Water governance can be traced back more than 5,000 years. However, it was only in 1982 and 1997, respectively, that global water agreements like the UN conventions on the seas and watercourses were adopted. Only recently has water become prominent on the global political agenda—for example, with the Mar del Plata conference of 1977, the Dublin conference on water in 1992, the water chapter in Agenda 21 adopted in 1992, the four World Water Forums since 1997, and the Millennium Declaration of 2000. Likewise, scientific work on water has only recently been globalised—for example, with the Global Water System Project of the Earth System Science Partnership and the UN-wide World Water Assessment Programme. Growing political interest in water research and governance raises the questions: What are the appropriate levels at which research and policy on water governance should be undertaken; and which issues should be addressed at which levels?

In this chapter, we argue that there are different levels at which water scholars and policymakers advocate governance, but the global perspective on water governance needs to be given more importance. We present a conceptual framework for analysing governance and conclude that present global water governance (GWG) is a mobius-web form of governance. Regarding the future of GWG, we identify four possible trends in line with scenarios prepared by scholars in related fields. We also consider the implications for water management of these different scenarios.

Why should global governance scholars be interested in GWG? To begin with, huge amounts of financial, administrative, and intellectual resources are being spent in the area of water governance and in related fields of environmental governance. A number of new ethical, conceptual, and management approaches are being generated and debated in different political contexts about how water resources can best be managed. These developments need to be recorded and assessed.

Water is a major global public good. Global governance theories that do not actively study the management of global public goods—particularly in the environmental area—will miss the rapidly evolving key issues of the twenty-first century. The contemporary world is extremely competitive, increasingly resource-scarce, and subject to far-reaching environmental change. As systemic and cumulative trends indicate global stress in the very resource base of society, including water, there are increasing calls for sustainability science and Earth System Governance theories. Studies of GWG can help to explore these possibilities.

Indeed, research on GWG and comparisons between GWG and other areas of environmental governance may provide valuable insights that could enrich the traditional core of global governance theories. For example, there is considerable discussion in the socioecological literature on the need to find transitional governance approaches to help society move from current unsustainable governance paradigms to future more sustainable governance paradigms. Notions of adaptive management are being increasingly promoted in the fields of water management and climate change. Perhaps such ideas could find application in wider areas of global governance such as economy, human rights, and security.

Water also provides a context for exploring the problematic of multilevel governance. Much as with other issue areas, researchers on water policy have tended to focus on one level only: the local, or the national, or the river basin. However, studies on water governance must adopt a multilevel approach to do justice to the complexity of current governance processes and challenges. Finally, global governance of water provides an occasion for wide-ranging multidisciplinary research. This issue requires collaboration not only among natural scientists, among engineers, and among social scientists, but also across these domains. To date, natural scientists and engineers are providing policy recommendations regarding water management, but this advice so far generally lacks input from global governance academics about how best to translate scientific findings into public policy. This gap needs to be filled.

The Need for Global Water Governance

Traditionally, resource governance has essentially been seen as a local-level issue. Many anthropologists and others continue to argue that one needs to understand local rights, needs, and stakeholders in order effectively to address governance issues. The driving forces behind this concept are the notions of decentralisation and subsidiarity. Since water problems are local, so goes this reasoning, they should be handled at the lowest appropriate governance level.

Another strong school of thought holds that water governance should be regulated at a national level. From this perspective, water is a national resource that should be governed for the benefit of the national economy and society: domestic interests come first. Increasingly, this vision is under threat, as many question the basis of the state and its willingness to promote the welfare of the people within it, and the need for international equity. A third approach to governance of water focuses on the basin level. This view argues that water-related problems and conflicts are best dealt with within the natural sphere of the system—that is, the hydrologically defined basin, catchment, or watershed.

This concept combines notions of efficiency with a hydrological systems approach. It allows for comprehensive problem analysis and the internalisation of otherwise externalized problems as they arise, for instance, from up- and downstream relations. Such research also studies the equitable management of transboundary or international waters. A fourth and relatively new school takes a global perspective on governance of water. Many water-related environmental and societal problems as well as water use-related conflicts elude appropriate solutions at the local level or within national or basin boundaries. In these cases, it is important to address issues at a global level. Thus, growing attention is being given to multilateralism in the international politics of water and to the recognition that local, national, and basin-level water issues are interlinked within a global water system.

These approaches—local, national, basin, and global—are not mutually exclusive. They indicate that different water issues are dealt with at different levels and that historically different perspectives have dominated. However, the present chapter and special issue highlight the global level as the relatively least understood and least explored aspect of water governance. Four arguments underscore the need for a global perspective alongside attention to local, national, and basin dimensions. First, the hydrological system is a global system, and exchange processes occur at the global level over relevant time periods. Examples include climate change impacts and teleconnections (patterns/phenomena that are related to each other through cause-effect relations across vast distances) between deforestation and precipitation.

Second, global environmental change and socioeconomic phenomena at the global level increasingly create situations in which the driving forces behind water-related problems and conflicts lie outside the reach of local, national, or basin-oriented governance regimes. Global trade impacts on water provide one such example. Third, many local environmental and social phenomena surrounding water are situated in global dynamics—for instance, of erosion, eutrophication, urbanisation, and biodiversity loss. Such local phenomena may, cumulatively, imply alarming global trends. For example, the construction of dams leads to a fragmentation and flow alteration of the world's river basins, with major and sometimes irreversible impacts on associated freshwater ecosystems. Fourth, many direct and indirect impacts of reductions in quantities and qualities of water are likely to be global in character (e.g., changed patterns of food production and bird migration). Each argument is linked with a specific kind of phenomena transforming the contemporary global water system and thus also with a specific kind of policy response. Phenomena occurring in the context of the first argument can only be dealt with efficiently through governance at the global level due to the physical nature of cause-effect relations; however, adaptation to climate change-induced floods or droughts, for example, needs to take place not only at but also within basins, and actions toward mitigating the indirect causes can only be taken at global levels, which then cascade down to national and local levels. Whereas mitigation of impacts in the context of the third argument (globally occurring local phenomena) lies within the range of local- or basin-level governance, finding efficient and effective solutions to problems and conflicts as well as approaching the driving forces call for a greater global understanding of and better coordinated activities on issues.

Having argued that there is a need for GWG, we must also emphasize those global arrangements for water cannot be understood without taking cross-level interactions into account. Water governance has a multilevel character, and global mechanisms must be incorporated in ways that are complementary to instruments applied at other levels. While some research has analysed global institutions in the field of water governance, global-and basin-level issues are often treated separately, without considering interplay across levels.

Thus, GWG can be defined as the development and implementation of norms, principles, rules, incentives, informative tools, and infrastructure to promote a change in the behaviour of actors at the global level in the area of water governance. Such governance calls for: (1) recognition that water has a global dimension; (2) inclusion of a global perspective on water governance at all other governance levels; and (3) adoption of more dense systems of multilevel governance from global to local level in the water field, because water policies—if they are to be effective—have to be simultaneously designed at global, fluvial, national, provincial, and local levels.

A Conceptual Framework for Analysing GWG

Governance can be viewed as either a normative or an analytical concept for policy research. Normatively, "good" governance relates to a regulatory system that shows qualities of accountability, transparency, legitimacy, public participation, justice, efficiency, the rule of law, and an absence of corruption. In contrast, an analytical approach provides a scientific basis for developing sound policy recommendations. There is an urgent need for analytical assessment of ongoing policy processes and their ability to meet future policy challenges. In particular, the following questions need to be addressed: How well can the present state of GWG deal with current policy challenges; and what future scenarios of GWG could deal effectively with emerging policy challenges?

Recent decades have brought a shift in social-scientific discourse from government to governance. This change in terminology signals a recognition that government is no longer the single decisionmaking authority, with sovereign control over the people and groups making up civil society. Instead, governance involves a multilevel, polycentric condition where many actors in different institutional settings contribute to policy development and implementation. Governance takes into account various governmental and nongovernmental actors and networks that together formulate and implement contemporary public policy. Governance encompasses coordination and steering processes involving formal as well as informal institutions. A major challenge is to understand how all these different processes in complex interrelations with one another determine policy outcomes and how change in governance regimes occurs.

With the complexity of governance systems, change results from a combination of formal regulations and informal self-organising processes among a range of actors. Although the governance concept has the strength of encompassing the complexity of contemporary policy processes, it at the same time presents a major challenge of encapsulating this complexity within a concise typological framework. One attempt at classification distinguishes between hierarchies, networks, and markets as three principal modes of governance. Thus, the dominating influence in a given governance arrangement may come from governmental control, from specialised networks, or from market-based structures.

Or the three modes may intertwine; it is increasingly realised that markets and networks function best within an official regulatory framework where contracts and the rule of law are respected. The development of GWG is influenced by (1) the tensions between globalisation and regionalisation; (2) the dominance of centralisation or decentralisation; (3) the diversity of formal and informal processes and outcomes; and (4) the influence of state versus nonstate actors and processes.

These dimensions are reflected in the governance typology of James Rosenau, which focuses on the structures and processes that sustain the flow of authority, whether they are in the form of commands or of requests for compliance.

The structural attribute distinguishes between either formal or informal frameworks, as well as a combination of formal and informal structures. The process attribute makes a distinction in terms of the flow of authority in a single direction (up or down) or in multiple directions (both up and down vertically, as well as both back and forth horizontally). This typology takes into account that the common linear models of governance (those in the left-hand column) have to be complemented with models taking into account nonlinear feedback and network processes.

These ideal types highlight distinctions between extremes and emphasize the dominant pathways of policy intervention. Governance includes different processes to varying degrees. The Rosenau typology is helpful in classifying actors and cross-level interactions. It also avoids normative claims and provides the basis for a strictly analytical framework.

However, the typology does not offer explicit guidance regarding the relative importance of state versus nonstate actors beyond asserting that states do not play an exclusive role in governance. Likewise, Rosenau's framework does not specify the relative significance of formality versus informality, of centralisation versus decentralisation, or of globalisation versus regionalisation. These matters must be assessed separately in each empirical case.

The Current State of GWG

The characteristics of governance regimes introduced in the preceding section can now be applied to analyse players and policy processes in the current GWG landscape. The discussion that follows identifies who is involved in policymaking around water—states or predominantly nonstate actors. The analysis further considers how far GWG is formal or informal, global or regional, centralised or decentralised. The results of GWG policy processes are also examined, in terms of binding agreements versus nonbinding declarations. Important aspects of contemporary GWG come in the form of international law.

These agreements generally reflect state practice and result from negotiations between states based on strict rules of procedure. They are relatively top-down, centralised global governance arrangements that lack broad involvement of stakeholder groups at different levels. GWG in this traditional statist vein includes the 1982 UN Convention on the Law of the Sea and its subsequent follow-up agreements, which aim to harmonize rules on managing the seas. Another statist construction, the 1997 UN Convention on the Law of the Non-navigational Uses of International Watercourses, aims to develop common norms for the management of watercourses. Although the Watercourses Convention has not entered into force, it has influenced policy regarding many international river basins. Other intergovernmental treaties address specific species (e.g., the Whaling Convention, 1946) and pollution of the seas (e.g., the convention related to dumping from nuclear ships, 1972).

Alongside international law, GWG also encompasses a number of permanent global intergovernmental agencies, particularly through the United Nations system. The UN is a relatively top-down organisation, although it does not exclude participation by nonstate actors and does not speak with one voice. Global governance of water involves a number of UN institutions (e.g., the Food and Agriculture Organisation [FAO], the UN Educational, Scientific and Cultural Organisation [UNESCO], the World Health Organisation [WHO], and the World Bank); funds (e.g., the Global Environment Facility [GEF]); and programs (e.g., the UN Development Programme [UNDP] and the UN Environment Programme [UNEP]).

In 2003, UN Water was established as an umbrella mechanism to coordinate UN implementation of the plan of action agreed upon at the 2002 World Summit on Sustainable Development (WSSD) and the Millennium Development Goals (MDGs), which includes reducing by half the proportion of people without sustainable access to safe drinking water and basic sanitation. UN Water endorses activities like the UNESCO-led World Water Assessment Programme (WWAP), which has since 2003 published a triennial World Water Development Report (WWDR).

In 2005, the UN launched the Water for Life Decade (2005-2015) to promote the achievement of the MDGs. Other activities fall fully under the aegis of one UN body (e.g., UNESCO's International Hydrological Programme [IHP]) or are jointly led by two or more bodies (e.g., the Global International Waters Assessment overseen by UNEP and GEF). Other aspects of GWG are organised at the regional level. In the European Union (EU), for example, the European Union Water Initiative (EUWI), launched at the 2002 Johannesburg summit, affirmed, among other things, that the EU would help achieve the MDG 7 and promote national water resources management plans by 2005. The EUWI is not a statist construction, however, as it relies largely on voluntary agreements for the mobilisation of financial and other resources from nonstate actors. Thus, EUWI is on the one hand a formal, centralised process and on the other hand an informal arrangement where state actors function side by side with market actors.

The mix of stakeholders in GWG is taken still further in the World Water Council (WWC), established in 1996. The WWC is an international multistakeholder platform that was established to address critical global water issues through, for example, a World Water Forum, held every three years since 1997.

The forum involves an intricate mobius web of interactions between state and nonstate actors and networks. Meanwhile the forums attract more than 15,000 participants, ranging from high-level ministerial delegations to local level managers and decisionmakers, from international water industry companies to local farmer irrigation boards, and include a wide range of nongovernmental organisations (NGOs), lobbyists, advocates, and researchers in such diverse fields as drinking water and sanitation, development aid, nature conservation, food security, and energy supply.

Linked to the forum is a ministerial conference that adopts a political declaration. The conference is an informal process with informal outcomes that are nonbinding and arguably ineffective. The proposed alternative of smaller, focused UN negotiations might be more effective, but they might arguably also be less legitimate, involving fewer nonstate actors.

Other aspects of GWG go further in the direction of private governance. For example, transnational water corporations (e.g., Suez, Ondeo, Veolia) are key global actors in water supply and wastewater treatment. About 10 percent of former public water services are now in private hands concentrated in urban centres. The privatization trend has supported the emergence of water companies, and the globalisation trend has promoted the dominance of global players who are able to outcompete smaller companies. The increasing importance of private companies and their self-governance initiatives is a case of "market governance."

Finally, global communities of scientists and water professionals increasingly shape GWG. The International Law Association (ILA) has been active on water matters since 1873, and in the 1970s UNESCO set up the International Hydrological Programme. At present, the professional community around water includes the Global Water Partnership, the International Water Association, the International Water Resources Association, the International Network of Basin Organisations, the Global Water System Project, the UNESCO-IHE Institute for Water Education, and many others. However, in contrast to other fields of research—such as climate change—professional circles around water do not constitute a well-defined epistemic community with a formal political mandate, nor do the professionals feed advice through formal channels into the governance process.

The preceding overview reveals the diffuse, heterogeneous, and fragmented character of today's GWG. A number of top-down, bottom-up, network, and side-by-side governance elements exist in parallel. The overall picture can be characterised as one of mobius web-type governance. Some initiatives are centralised, while others are decentralised. Some have a strongly global character, while others tend to be predominantly regional. No centralised UN agency is authorised to make policy, and UN Water as a platform for collaboration among UN agencies does not have a strong mandate. There are no indications of an emergent global leadership. Water is thus a rising issue of global governance characterised by comparatively young and immature structures and processes that have slowly evolved over the past two decades.

This immaturity is strange given that global governance has developed relatively more rapidly in other policy fields, such as the depletion of the ozone layer and climate change. There, formal links with organised epistemic communities have been arranged. This fuller development may have occurred because ozone depletion and climate change were both from the start perceived to be global problems that call for global governance. These challenges could not be addressed unless everyone participated in one way or other. Moreover, both problems called for the reduction of specific emissions—namely, ozone-depleting substances and greenhouse gases. Relative clarity of focus made it easier to generate concentrated governance programs. At the same time, mitigation strategies could focus on relatively simple technological solutions—for example, the phaseout of ozone-depleting substances and the reduction of [CO_2] emissions.

Obviously, this characterisation oversimplifies both of these cases of global governance. The actual picture is more complex, and these complexities have made climate governance increasingly more diffuse at the global level. Nevertheless, water is different from ozone depletion and climate change. For example, water issues are "creeping problems," having developed into global problems slowly over the centuries, whereas ozone and climate questions have become acute in a matter of decades. Moreover, the global character of water problems is not always evident: local water problems can be addressed locally. Although up to 2 billion people have poor access to potable water and sanitation services, this problem is not seen as dramatic. Nor are matters of water distribution susceptible to "easy" technological solutions. Awareness of the global dimension of water-related problems and the need for global governance processes to address them is only slowly emerging. In addition, the cross-sectoral character of water (across developmental, environmental, economic, and security dimensions) hampers the development of a well-structured and clear-cut global governance system.

Future Scenarios for GWG

As seen in the preceding discussion, water is an emergent problem of global public policy whose governance is so far characterised by a mobius-web structure. The various global initiatives developed to date appear to compete for influence rather than move toward coordination. This competition may be appropriate and fruitful in an early stage of developing governance.

However, for the longer-term future it becomes necessary to consider different arrangements and their potential for dealing with increasingly urgent challenges of sustainable global management of water resources. To this end the following paragraphs elaborate scenarios that are alternative projections of the future of GWG. Scenarios are a useful way to understand how driving forces may shape future developments, to assess the associated uncertainties, and to prepare for these circumstances.

Based on current trends, four possible future directions for the development of GWG can be identified. These include (1) global policymaking through informal, decentralised, and market approaches; (2) global policymaking with a focus on formal treaties and centralised agencies; (3) regional, multilevel policymaking; and (4) state-centred regional policymaking. It should be noted here that these are stylized ideal-typical scenarios that cannot take into account all nuances of possible developments. Clearly there may be informal networks that are centralised and formal bodies that are decentralised. The four scenarios are explained in the following paragraphs.

In scenario 1, the future GWG apparatus would be global, informal, decentralised, and privatized, involving multiple competing or complementary multilevel initiatives. This scenario would take forward trends since the 1990s of multiple global initiatives, mostly undertaken by nonstate actors and the UN system. The resulting situation would be one of fragmented governance with partly competing and partly complementary platforms and initiatives. One can conceive of a growing competition between such initiatives (e.g., the World Water Forums, World Water Week, the World Water Council, the International Water Association, UN Water, and the ILA), without the dominance of one process or actor. Alternatively, one (formal) actor could take the lead and support collaboration between complementary efforts. However, such collaboration is unlikely to emerge without leadership. Scenario 2 also envisions highly globalised governance, but with an emphasis on formal treaty-based regimes. Apart from the UN conventions on the sea and on watercourses, other relevant global treaties cover a range of issues like climate change and trade. Increasing pressures on water resources and increasing unhappiness with the diffuse and unclear nature of mobius web-like approaches could raise the pressure to develop more formal treaty-based global governance.

Scenario 3 would shift the centre of gravity in GWG from the global to the regional level, albeit in a context of continuing to develop multilevel arrangements. In a multispeed world, where parts of the globe have advanced systems of regional cooperation while others lack them, some parties may prefer to regulate water issues at a regional level. Regionalists may fear that agreements at the global level do not go far enough or that global arrangements may neglect region-specific contexts. An example of a regional, multilevel agreement is the EU, with its harmonized joint legislation on water issues (e.g., the Water Framework Directive).

Scenario 4 would likewise see more region-centred GWG, but this time with greater initiative centred in states. In effect, this would mean returning GWG to the pattern that prevailed before the 1990s, when states would negotiate treaties with neighbouring states on the basis of narrow national interests. While a trend since the 1990s has seen moves to consolidate earlier arrangements in global agreements, the rise of terrorism and the return to unilateralism by the United States, as well as fears of the impact of climate change on national water systems, could take the focus back to state-centred management of water issues. None of these four scenarios for future GWG is by itself sufficient to address global water issues. A combination of global informal networks and markets (scenario 1) and a global treaty approach (scenario 2) may have the potential to do so, since it combines global coordination with flexibility and adaptive negotiation spaces. Formal regulatory approaches at the global level (scenario 3) tend to remain rather vague and face tremendous difficulties in triggering real action, as the experience with the implementation of UN conventions in the water sectors or the MDGs show. However, they create valuable political commitment as the basis for further multilateral and regional activities.

State-centred regional activities (scenario 4) are likely to continue playing an important role in water governance but are not suitable for addressing the GWG challenges at hand. To embed these scenarios for GWG in a wider context, they can be linked to other scenario work undertaken by the Intergovernmental Panel on Climate Change (IPCC) in 2000, by the Millennium Assessment (MA) in 2005, and by the fourth Global Environment Outlook (GEO) in 2007.

The IPCC developed four scenarios, named A1, A2, B1, and B2. The A1 scenario is broadly in line with GWG scenario 1. It assumes a future with rapid economic growth, rapid technological change, a peak in global population in the middle of the twenty-first century, and increased cultural and social convergence among regions. The IPCC's A2 scenario visualizes a heterogeneous world where countries and people focus on self-reliance and local identities, rather on the lines of GWG scenario 4. In IPCC's B1 world, regions converge toward a service and information society, there is much focus on environmental impacts, and governance promotes sustainability and equity. The emphasis on global solutions broadly mirrors GWG scenario 2. The B2 scenario focuses on decentralised solutions emphasizing local routes to sustainability, which again goes in a similar direction as GWG scenario 4.

The MA also developed four scenarios for the future. One, dubbed Global Orchestration, is broadly in congruence with GWG scenario 1. It focuses on sustainable development and fair trade, with enhancement of global public goods and global education. A second MA scenario, called Order from Strength, shows significant similarities with GWG scenario 4. It focuses on conservation efforts such as reserves, regional trade blocs, security, and protection and is highly regional in character. The third MA scenario, labelled Adapting Mosaic, largely corresponds to GWG scenario 3.

It assumes a focus on local and regional comanagement owing to discredited global institutions in a rather fragmented world. The fourth MA scenario, Techno Garden, shows considerable overlap with GWG scenario 1, with its market-oriented and informal character. It emphasizes green technology, tradable rights, free movement of goods, technical expertise, and the like.

The GEO of 2007 explores four scenarios in a somewhat different manner. The scenarios mainly distinguish between different types of interactions between state and nonstate actors and their performance on environment and sustainability issues. The scenario dubbed Policy First relies on top-down approaches to environmental governance. Here, government implements strong policies to improve environmental and human well-being with the support of private and civil society actors.

This scenario has similarities with GWG scenarios 2 and 4. The GEO scenario Sustainability First has government, civil society, and the private sector collaborating to improve environmental and human well-being, with a strong emphasis on equity. With its implicit reliance on informal and decentralised processes, this GEO scenario has similarities with GWG scenarios 1 and 3. A third GEO scenario, called Security First, focuses on national and regional actors as is the case in GWG scenarios 3 and 4. It envisions government and the private sector competing for control in efforts to improve, or at least maintain, human well-being, mainly for the rich and powerful in society.

Finally, the GEO scenario called Markets First, emphasizes decentralised and informal market mechanisms at the expense of centralised and formalised policy interventions and therefore overlaps with GWG scenarios 1 and 3. Here the private sector, with active government support, pursues maximum economic growth as the best path to improve environmental and human well-being.

Like the GWG scenarios, the IPCC and MA scenarios (but not the GEO scenarios) make more or less clear distinctions between the four opposing trends that were identified earlier in this chapter: global versus regional, centralised versus decentralised, formal versus informal, and state versus non-state. The comparison of scenarios shows that the exploratory analysis of GWG scenarios is generally in line with other projections of future global environmental governance. The comparison also shows that global scenarios differ in the kind of emphasis they give to different elements of governance and the kind of development they consider to be plausible. However, further research may reveal more inconsistencies. Perhaps scholars working on scenarios for global environmental governance (including in respect of water) should strive for a more consistent terminology. Such a terminology should not reduce the diversity of approaches but rather allow comparing why and how they differ.

It is unclear how actual future GWG will develop in relation to the different scenarios. What is clear is that a number of activities are taking place on a number of fronts. Each has advantages and disadvantages. A state-centred approach appears to be outdated, but given the crucial importance of water as a politically sensitive resource, many governments will be reluctant to cede control to fluvial or global-level regimes. Yet a state-centred multilateral regime may inhibit the development of cooperative norms and instruments for the global management of water. Since world regions are in different stages of development, it is likely that water will be treated more as a regional issue and that each region will address the problem within its existing capabilities. A pluralist, global, multiactor approach may work well in the developed countries, where institutional processes exist to ensure accountability of the key actors, but there are major doubts about whether such processes can function well in the developing countries.

In this chapter, we have developed a conceptual framework for analysing global governance of water. Our analysis of GWG, as well as the development of future projections, has identified clear challenges for global governance scholars. Emphasis needs to be given to cross-level interactions, the development of consistent terminologies, and the construction of a framework that allows better comparative analysis. Our analysis of present circumstances concludes that GWG is currently diffuse and mobius web-like in character. A lack of strong motivation on the part of UN agencies and states to push water management has encouraged the rise of pluralistic bodies that try to deal with these issues. However, it is not clear that these polycentric governance frameworks can be more successful in generating the necessary political will for global action. Some kind of formal global coordination is required in tandem with more decentralised network and market-based approaches.

However, further questions remain that lie beyond the scope of this chapter. For example, can effective and legitimate global governance systems for water be developed in a world of decreasing social solidarity and increasing economic and environmental stress? What kind of frameworks for global governance of water should be advocated from a normative perspective, as against the purely analytical approach maintained in the present discussion? Hopefully, the theoretical exploration we have undertaken will stimulate further research along these and other lines into this vital question of global resource management.

9

Surface-water Hydrology

Surface-water hydrology is a field that encompasses all surface waters of the globe (overland flows, rivers, lakes, wetlands, estuaries, oceans, etc). This a subset of the hydrologic cycle that does not include atmospheric, and ground waters. Surface-water hydrology relates the dynamics of flow in surface-water systems (rivers, canals, streams, lakes, ponds, wetlands, marshes, arroyos, oceans, etc.). This includes the field measurement of flow (discharge); the statistical variability at each setting; floods; drought susceptibility and the development of the levels of risk; and the fluid mechanics of surface waters.

In-depth analysis of surface-water components of the hydrologic cycle: hydrometeorology, evaporation/transpiration, rainfall-runoff relationships, open-channel flow, flood hydrology, fluid mechanics, and statistical and probabilistic methods in hydrology. Surface-water hydrology includes the relation between rainfall and surface runoff; this relationship is an important aspect of water resources for sewerage (wastewater) or (sewage), drinking water, agriculture (irrigation) environmental protection, and for flood control.

The relationships between groundwater and surface water includes baseflow needs for instream flow, and subsurface water levels in wells. A stormwater engineer is a civil engineer who manages the flow, filtering, and release of stormwater.

Drainage Basin

A drainage basin is an extent or an area of land where surface water from rain and melting snow or ice converges to a single point, usually the exit of the basin, where the waters join another waterbody, such as a river, lake, reservoir, estuary, wetland, sea, or ocean. In closed drainage basins the water converges to a single point inside the basin, known as a sink, which may be a permanent lake, dry lake, or a point where surface water is lost underground. The drainage basin includes both the streams and rivers that convey the water as well as the land surfaces from which water drains into those channels, and is separated from adjacent basins by a drainage divide. The drainage basin acts as a funnel by collecting all the water within the area covered by the basin and channelling it to a single point. Each drainage basin is separated topographically from adjacent basins by a geographical barrier such as a ridge, hill or mountain. Other terms that are used to describe a drainage basin are catchment, catchment area, catchment basin, drainage area, river basin, water basin and watershed. In the technical sense, a watershed refers to a divide that separates one drainage area from another drainage area. However, in the United States and Canada, the term is often used to mean a drainage basin or catchment area itself. Drainage basins drain into other drainage basins in a hierarchical pattern, with smaller sub-drainage basins combining into larger drainage basins. Drainage basins are similar but not identical to hydrologic units, which are drainage areas delineated so as to nest into a multi-level hierarchical drainage system. Hydrologic units are designed to allow multiple inlets, outlets, or sinks. In a strict sense, all drainage basins are hydrologic units but not all hydrologic units are drainage basins.

The United States Environmental Protection Agency launched the website Watershed Central for the US public to exchange information and locate resources needed to restore local drainage basins in that country.

Major Drainage Basins of the World

Ocean Basins

There are 354 drainage basins throughout the world in all sorts of places. The following is a list of some of the major ones:

- About 47% of all land in the world drains to the Atlantic Ocean. In North America, surface water drains to the Atlantic via the Saint Lawrence River and Great Lakes basins, the Eastern Seaboard of the United States, the Canadian Maritimes, and most of Newfoundland and Labrador. Nearly all of South America east of the Andes also drains to the Atlantic, as does most of Western and Central Europe, and the greatest portion of western Sub-Saharan Africa. The three major mediterranean seas of the world also flow to the Atlantic:
  - — The American Mediterranean Sea (the Caribbean Sea and Gulf of Mexico) basin includes most of the American interior between the Appalachian and Rocky Mountain ranges, a small part of the Canadian provinces of Alberta and Saskatchewan, eastern Central

America, the islands of the Caribbean and the Gulf, and a small part of northern South America.

  - The European Mediterranean Sea basin includes much of northern Africa, east-central Africa (through the Nile), southern, central, and eastern Europe, Turkey, and the coastal areas of Israel, Lebanon, and Syria.
  - The Arctic Ocean basin drains most of Western and Northern Canada east of the Continental Divide, the north shore of Alaska and parts of North Dakota, South Dakota, Minnesota, and Montana in the United States, the north shore of the Scandinavian peninsula in Europe, and much of central and northern Russia.

- Just over 13% of the land in the world drains to the Pacific Ocean. Its basin includes much of China, southeastern Russia, Japan, Korea, most of Indonesia and Malaysia, the Philippines, all of the Pacific Islands, the northeast coast of Australia, and the western parts of Canada, the United States (including most of Alaska), Central America, and South America.
- The Indian Ocean's drainage basin also comprises about 13% of Earth's land. It drains the eastern coast of Africa, the coasts of the Red Sea and the Persian Gulf, the Indian subcontinent, Burma, and most of Australia.
- The Southern Ocean drains Antarctica. Antarctica comprises approximately eight percent of the Earth's land.

Largest River Basins

The three largest river basins (by area), from largest to smallest, are the Amazon basin, the Congo basin, and the Mississippi basin. The three rivers that drain the most water, from most to least, are the Amazon, Congo, and Ganges Rivers.

Endorheic Drainage Basins

Endorheic drainage basins are inland basins that do not drain to an ocean. Around 18 per cent of all land drains to endorheic lakes or seas or sinks. The largest of these consists of much of the interior of Asia, and drains into the Caspian Sea and the Aral Sea. Other endorheic regions include the Great Basin in the United States, much of the Sahara Desert, the watershed of the Okavango River (Kalahari Basin), highlands near the African Great Lakes, the interiors of Australia and the Arabian Peninsula, and parts in Mexico and the Andes. Some of these, such as the Great Basin, are not single drainage basins but collections of separate, adjacent closed basins.

In endorheic bodies of standing water where evaporation is the primary means of water loss, the water is typically more saline than the oceans. An extreme example is the Dead Sea.

Importance of Drainage Basins

Geopolitical Boundaries

Drainage basins have been historically important for determining territorial boundaries, particularly in regions where trade by water has been important. For example, the English crown gave the Hudson's Bay Company a monopoly on the fur trade in the entire Hudson Bay basin, an area called Rupert's Land. Today, bioregional democracy can include agreements of states in a particular drainage basin to defend it. One example of this is the Great Lakes Commission.

Hydrology

In hydrology, the drainage basin is a logical unit of focus for studying the movement of water within the hydrological cycle, because the majority of water that discharges from the basin outlet originated as precipitation falling on the basin. A portion of the water that enters the groundwater system beneath the drainage basin may flow towards the outlet of another drainage basin because groundwater flow directions do not always match those of their overlying drainage network. Measurement of the discharge of water from a basin may be made by a stream gauge located at the basin's outlet.

Rain gauge data is used to measure total precipitation over a drainage basin, and there are different ways to interpret that data. If the gauges are many and evenly distributed over an area of uniform precipitation, using the arithmetic mean method will give good results. In the Thiessen polygon method, the watershed is divided into polygons with the rain gauge in the middle of each polygon assumed to be representative for the rainfall on the area of land included in its polygon. These polygons are made by drawing lines between gauges, then making perpendicular bisectors of those lines form the polygons. The isohyetal method involves contours of equal precipitation are drawn over the gauges on a map. Calculating the area between these curves and adding up the volume of water is time consuming.

Geomorphology

Drainage basins are the principal hydrologic unit considered in fluvial geomorphology. A drainage basin is the source for water and sediment that moves through the river system and reshapes the channel.

Ecology

Drainage basins are important elements to consider also in ecology. As water flows over the ground and along rivers it can pick up nutrients, sediment, and pollutants. Like the water, they get transported towards the outlet of the basin, and can affect the ecological processes along the way as well as in the receiving water source. Modern usage of artificial fertilizers, containing nitrogen, phosphorus, and potassium, has affected the mouths of watersheds. The minerals will be carried by the watershed to the mouth and accumulate there, disturbing the natural mineral balance. This can cause eutrophication where plant growth is accelerated by the additional material.

Resource Management

Because drainage basins are coherent entities in a hydrological sense, it has become common to manage water resources on the basis of individual basins. In the U.S. state of Minnesota, governmental entities that perform this function are called watershed districts. In New Zealand, they are called catchment boards. Comparable community groups based in Ontario, Canada, are called conservation authorities. In North America this function is referred to as watershed management. In Brazil, the National Policy of Water Resources, regulated by Act n° 9.433 of 1997, establishes the drainage basin as territorial division of Brazilian water management.

Catchment Factors

The catchment is the most significant factor determining the amount or likelihood of flooding. Catchment factors are: topography, shape, size, soil type and land use (paved or roofed areas). Catchment topography and shape determine the time taken for rain to reach the river, while catchment size, soil type and development determine the amount of water to reach the river.

Topography

Topography determines the speed with which the runoff will reach a river. Clearly rain that falls in steep mountainous areas will reach the river faster than flat or gently sloping areas.

Shape

Shape will contribute to the speed with which the runoff reaches a river. A long thin catchment will take longer to drain than a circular catchment.

Size

Size will help determine the amount of water reaching the river, as the larger the catchment the greater the potential for flooding.

Soil Type

Soil type will help determine how much water reaches the river. Certain soil types such as sandy soils are very free draining and rainfall on sandy soil is likely to be absorbed by the ground. However, soils containing clay can be almost impermeable and therefore rainfall on clay soils will run off and contribute to flood volumes. After prolonged rainfall even free draining soils can become saturated, meaning that any further rainfall will reach the river rather than being absorbed by the ground.

Land Use

Land use can contribute to the volume of water reaching the river, in a similar way to clay soils. For example, rainfall on roofs, pavements and roads will be collected by rivers with almost no absorption into the groundwater.

Water Distribution on Earth

Water is widely distributed on Earth as freshwater and salt water in the oceans. The Earth is often referred to as the "blue planet" because when viewed from space it appears blue. This blue colour is caused by reflection from the oceans which cover roughly 71 per cent of the area of the Earth.

The oceanic crust is young, thin and dense, with none of the rocks within it dating from any older than the break-up of Pangaea. Because water is much denser than any gas, this means that water will flow into the "depressions" formed as a result of the high density of oceanic crust.

(On a planet like Venus, with no water, the depressions appear to form a vast plain above which rise plateaux). Since the low density rocks of the continental crust contain large quantities of easily eroded salts of the alkali and alkaline earth metals, salt has, over billions of years, accumulated in the oceans as a result of evaporation returning the fresh water to land as rain and snow.

As a result, the vast bulk of the water on Earth is regarded as saline or salt water, with an average salinity of 35‰ (or 3.5 per cent, roughly equivalent to 35 grams of salts in 1kg of seawater), though this varies slightly according to the amount of runoff received from surrounding land.

In all, oceanic water, saline water from marginal seas, and water from saline *closed* lakes amounts to over 98 per cent of the water on Earth, though no closed lake stores a globally significant amount of water. Renewable *saline* groundwater is believed to total at least 100 km^3 globally, but is seldom considered except when evaluating water quality in arid regions.

The remainder of the Earth's water constitutes the planet's fresh water resource. Typically, fresh water is defined as water with a salinity of less than 1 per cent that of the oceans - i.e. below around 0.35‰. Water with a salinity between this level and 1‰ is typically referred to as marginal water because it is marginal for many uses by humans and animals.

The planet's fresh water is also very unevenly distributed. Although in warm periods such as the Mesozoic and Paleogene when there were no glaciers anywhere on the planet all fresh water was found in rivers and streams, today the distribution is approximately as follows:

- Ice caps and glaciers - 68.7%, of which
  - Antarctic ice cap - 90%, 9700 years renewal interval
  - Greenland ice cap - 9%
  - Other glaciers - <1%, 1600 years renewal interval
- Groundwater - 30.1%, 1400 year renewal interval
- Surface water - 0.3%, of which
  - Freshwater lakes - 87%, 17 years renewal interval
  - Swamps - 11%
  - Rivers - 2%, 16 days renewal interval
- Ground ice and permafrost - 0.86%
- Atmosphere 0.04%

Of these sources, only river water is generally valuable. Most water in lakes is in very inhospitable regions such as glacial lakes of Canada. Lake Baikal and Lake Khovsgol, both protected from Quaternary glaciation by aridity, have equivalent amounts of water, and the latter has been used in Mongolia as a source of drinking water.

Although the total volume of groundwater is known to be much greater than that of river runoff, a large proportion of this groundwater is saline and should therefore be classified with the saline water above. There is also a lot of *fossil* groundwater in arid regions that has never been renewed for thousands of years; this must not be seen as renewable water.

However, fresh groundwater is of great value, especially in arid countries such as India. Its distribution is broadly similar to that of surface river water, but it is easier to store in hot and dry climates because groundwater storages are much more shielded from evaporation than are dams. In countries such as Yemen, groundwater from erratic rainfall during the rainy season is the major source of irrigation water.

Because groundwater recharge is much more difficult to accurately measure than surface runoff, groundwater is not generally used in areas where even fairly limited levels of surface water are available. Even today, estimates of total groundwater recharge vary greatly for the same region depending on what source is used, and cases where fossil groundwater is exploited beyond the recharge rate (including the Ogallala Aquifer) are very frequent and almost always not seriously considered when they were first developed.

Distribution of River Water

The distribution of renewable river water across the Earth's surface is very uneven. Earth: 28% Water: 72%

Continent or region	***Renewable river water (km³)***	***Percent of world total***
Sub-Saharan Africa	4,000	9.20
Middle East and North Africa	140	0.32
Europe	2,900	6.70
Asia (excluding Middle East)	13,300	30.6
Australia	440	1.01
Oceania	6,500	14.9
North America	7,800	17.9
South America	12,000	27.6

Even within these regions, there can be huge variations. For example, as much as a quarter of Australia's limited renewable fresh water supply is found in almost uninhabited Cape York Peninsula.

Also, even in well-watered continents, there are areas that are extremely short of water, such as Texas in North America, whose renewable water supply totals only 26 km^3 in an area of 695,622 km^2, or South Africa, with only 44 km^3 in 1,221,037 km^2. The areas of greatest concentration of renewable water are:

- The Amazon and Orinoco Basins (a total of 6,500 km^3 or 15 percent of global runoff)
- East Asia
  - — Yangtze Basin - 1,000 km^3
- South and Southeast Asia, with a total of 8,000 km^3 or 18 percent of global runoff
  - — Brahmaputra Basin - 900 km^3
  - — Irrawaddy Basin - 500 km^3
  - — Mekong Basin - 450 km^3
- Canada, with over 10 percent of world's river water and large numbers in lakes
  - — Mackenzie River - over 250 km^3
  - — Yukon River - over 150 km^3
- Siberia
  - — Yenisey - over 5% of world's fresh water in basin - second largest after the Amazon
  - — Ob River - over 500 km^2
  - — Lena River - over 450 km^2

- New Guinea
  - Fly and Sepik Rivers - total over 300 km^3 in only about 150,000 km^2 of basin area.

And also the ratio of salt water to fresh is 40 salt to 1 fresh.

Variability of Water Availability

Variability of water availability is of major importance both for the functioning of aquatic species and also for the availability of water for human use: water that is only available in a few wet years must not be considered renewable. Because most global runoff comes from areas of very low climatic variability, the total global runoff is generally of low variability.

Indeed, even in most arid zones, there tends to be few problems with variability of runoff because most usable sources of water come from high mountain regions which provide highly reliable glacier melt as the chief source of water, which also comes in the summer peak period of high demand for water. This historically aided the development of many of the great civilizations of ancient history, and even today allows for agriculture in such productive areas as the San Joaquin Valley.

However, in Australia and Southern Africa the story is different. Here, runoff variability is much higher than in other continental regions of the world with similar climates. Typically temperate (Köppen climate classification C) and arid (Köppen climate classification B) climate rivers in Australia and Southern Africa have as much as three times the coefficient of variation of runoff of those in other continental regions.

The reason for this is that, whereas all other continents have had their soils largely shaped by Quaternary glaciation and mountain building, soils of Australia and Southern Africa have been largely unaltered since at least the early Cretaceous and generally since the previous ice age in the Carboniferous. Consequently available nutrient levels in Australian and Southern African soils tend to be orders of magnitude lower than those of similar climates in other continents, and native flora compensate for this through much higher rooting densities (e.g. proteoid roots) to absorb minimal phosphorus and other nutrients.

Because these roots absorb so much water, runoff in typical Australian and Southern African rivers does not occur until about 300mm (12 inches) or more of rainfall has occurred. In other continents, runoff will occur after quite light rainfall due to the low rooting densities.

The consequence of this is that many rivers in Australia and Southern Africa (as compared to *extremely few* in other continents) are theoretically impossible to regulate because rates of evaporation from dams mean a storage sufficiently large to theoretically regulate the river to a given level would actually allow very little draft to be used. Examples of such rivers include those in the Lake Eyre Basin. Even for other Australian rivers, a storage three times as large is needed to provide a third the supply of a comparable climate in southeastern North America or southern China. It also effects aquatic life, favouring strongly those species able to reproduce rapidly after high floods so that some will survive the next drought.

Climate type (Köppen)	Mean annual rainfall	Typical runoff ratio for Australia and Southern Africa	Typical runoff ratio for rest of the world
BWh	250mm (10 inches)	1 percent (2.5mm)	10 percent (25mm)
BSh (on Mediterranean fringe)	350mm (14 inches)	3 percent (12mm)	20 percent (80mm)
Csa	500mm (20 inches)	5 percent (25mm)	35 percent (175mm)
Caf	900mm (36 inches)	15 percent (150mm)	45 percent (400mm)
Cb	1100mm (43 inches)	25 percent (275mm)	70 percent (770mm)

Tropical (Köppen climate classification A) climate rivers in Australia and Southern Africa do not, in contrast, have markedly lower runoff ratios than those of similar climates in other regions of the world. Although soils in tropical Australia and southern Africa are even poorer than those of the arid and temperate parts of these continents, vegetation can use organic phosphorus or phosphate dissolved in rainwater as a source of the nutrient. In cooler and drier climates these two related sources tend to be virtually useless, which is why such specialised means are needed to extract the most minimal phosphorus.

There are other isolated areas of high runoff variability, though these are basically due to erratic rainfall rather than different hydrology. These include:

- Southwest Asia
- The Brazilian Nordeste
- The Great Plains of the United States

Atmospheric Escape

Atmospheric escape is the loss of planetary atmospheric gases to outer space. There are several processes that can cause this. In some cases, it can be a very important process; for example, both Venus and Mars have probably lost much of their water due to atmospheric escape since they have a weaker gravitational field strength than Earth.

Thermal Escape Mechanisms

One classical thermal escape mechanism is Jeans escape. In a quantity of gas, the average velocity of a molecule is determined by temperature, but the velocity of individual molecules varies continuously as they collide with one another, gaining and losing kinetic energy. The variation in kinetic energy among the molecules is described by the Maxwell distribution. The kinetic energy and mass of a molecule determine its velocity by $E_{kin} = \frac{1}{2}mv^2$.

Individual molecules in the high tail of the distribution may reach escape velocity, at a level in the atmosphere where the mean free path is comparable to the scale height, and leave the atmosphere. The more massive the molecule of a gas is, the lower the average velocity of molecules of that gas at a given temperature, and the less likely it is that any of them reach escape velocity. This is why hydrogen escapes from a given atmosphere more easily than carbon dioxide. Also, if the planet has a higher mass, the escape velocity is greater, and fewer particles will escape. This is why the gas giant planets have significant amounts of hydrogen and helium, which escape on Earth. Distance from the star also plays a part; a close planet has a hotter atmosphere, with a faster range of velocities, and more chance of escape. A distant body has a cooler atmosphere, with a slower range of velocities, and less chance of escape. This helps Titan, which is small compared to Earth but further from the Sun, keep its atmosphere. However, while it has not been observed, it is theorised that an atmosphere with a high enough pressure and temperature can undergo a 'blow-off'. In this situation molecules basically just flow off into space. Here it is possible to lose heavier molecules that would not normally be lost.

Significance of Solar Winds

The relative importance of each loss process is a function of planet mass, atmosphere composition, and distance from a star. A common erroneous belief is that the primary non-thermal escape mechanism is atmospheric stripping by a solar wind in the absence of a magnetosphere. Excess kinetic energy from solar winds can impart sufficient energy into atmospheric particles to reach escape velocity, causing atmospheric escape.

The solar wind, composed of ions, is deflected by magnetic fields because the charged particles within the wind flow along magnetic field lines. The presence of a magnetic field thus deflects solar winds, preventing atmospheric loss to solar winds. On Earth, for instance, the interaction between the solar wind and magnetic field deflects the solar wind around the planet, with near total deflection around 10 Earth radii away. This region of deflection is called a bow shock.

Depending on planet size and atmospheric composition, however, a lack of magnetic field does not determine the fate of a planet's atmosphere. Venus, for instance, has no powerful magnetic field. Its close proximity to the sun also increases the speed and number of particles, and would presumably cause the atmosphere to be stripped almost entirely, much like that of Mars. Despite this, the atmosphere of Venus is two orders of magnitudes denser than Earth's. Recent models indicate that stripping by solar wind accounts for less than 1/3 of total non-thermal loss processes.

While Venus and Mars have no magnetosphere to protect the atmosphere from solar winds, photoionising radiation (sunlight) and the interaction of the solar wind with the atmosphere of the planets causes ionisation of the uppermost part of the atmosphere. This ionised region of atmosphere, in turn, induces magnetic moments that deflect solar winds much like a magnetic field, limiting solar wind effects to the uppermost altitudes of atmosphere, roughly 1.2-1.5 planetary radii away from the planet, or an order of magnitude closer to the surface than Earth's magnetic field creates. Past this region, also called a bow shock, the solar wind is slowed to subsonic velocities. Nearer to the surface, solar wind dynamic pressure balances with pressure from the ionosphere, at a region called the ionopause. This interaction typically prevents solar wind stripping from being the dominant loss process of the atmosphere.

Comparison of Non-thermal Loss Processes Based on Planet and Particle Mass

Dominant non-thermal loss processes differ based on the planetary body in discussion. The varying relative significance of each process is based on planetary mass, atmospheric composition, and distance from the sun. The dominant non-thermal loss processes for Venus and Mars, two terrestrial planets without magnetic fields, are dissimilar. The dominant non-thermal loss process on Mars is pick-up from solar winds, because the atmosphere is not dense enough to shield itself from the winds during peak solar activity. Venus is somewhat shielded from solar winds by merit of a denser atmosphere, and solar pick-up is not the dominant non-thermal loss process on Venus. Smaller bodies without magnetic fields are more likely to suffer from solar winds, because the planet is too small to hold sufficient atmosphere to stop solar winds.

The dominant loss process for Venus is loss through electric force field acceleration. Because electrons are more mobile than other particles, they are more likely to escape from the top of the ionosphere of Venus. As a result, a minor net positive charge can develop. The net positive charge, in turn, creates an electric field that can accelerate other positive charges out of the system. Through this, H+ ions are accelerated beyond escape velocity, causing atmospheric escape through this process. Other important loss processes on Venus are photochemical reactions, driven by proximity to the Sun.

Photochemical reactions rely on splitting the molecules into constituent atoms, often with a significant portion of kinetic energy maintained in the less massive particle. This particle is of sufficiently low mass and high kinetic energy to escape from Venus. Oxygen, relative to hydrogen, is not of sufficiently low mass to escape through this mechanism on Venus.

Phenomena of Non-thermal Loss Processes on Moons with Atmospheres

Several moons within our solar system have atmospheres and are subject to atmospheric loss processes. They typically have no magnetic fields of their own, but orbit planets with powerful magnetic fields. Many of these moons lie within the magnetic fields generated by the planets and are less likely to undergo sputtering and pick-up. The shape of the bow shock, however, allows for some moons, such as Titan, to pass through the bow shock when their orbits takes them between the sun and their primary. Titan spends roughly half of its transit time outside of the bow-shock and being subjected to unimpeded solar winds. The kinetic energy gained from pick-up and sputtering associated with the solar winds increases thermal escape throughout the transit of Titan, causing neutral hydrogen to escape from the moon. The escaped hydrogen maintains an orbit following in the wake of Titan, creating a neutral hydrogen torus around Saturn. Io, in its transit around Jupiter, encounters a plasma cloud. Interaction with the plasma cloud induces sputtering, kicking off sodium particles. The interaction produces a stationary banana-shaped charged sodium cloud along a part of the orbit of Io.

Impact Erosion

The impact of a large meteoroid can lead to the loss of atmosphere. If a collision is energetic enough, it is possible for ejecta, including atmospheric molecules, to reach escape velocity. Just one impact such as the Chicxulub event does not lead to a significant loss, but the terrestrial planets went through enough impacts when they were forming for this to matter.

Sequestration

This is a loss, not an escape; it is when molecules solidify out of the atmosphere onto the surface. This happens on Earth, when water vapour forms glacial ice or when carbon dioxide is sequestered in sediments. The dry ice caps on Mars are also an example of this process.

One mechanism for sequestration is chemical; for example, most of the carbon dioxide of the Earth's original atmosphere has been chemically sequestered into carbonate rock. Very likely a similar process has occurred on Mars. Oxygen can be sequestered by oxidation of rocks; for example, by increasing the oxidation states of ferric rocks from Fe^{2+} to Fe^{3+}. Gases can also be sequestered by adsorption, where fine particles in the regolith capture gas which adheres to the surface of particles.

Dominant Atmospheric Escape and Loss Processes on Earth

Earth is too large to lose particles efficiently through Jeans Escape. The exosphere is the high altitude region where atmospheric density is sparse and Jeans Escape occurs. Jeans escape calculations assuming an exosphere temperature of 1,800 degrees show that to deplete O+ ions by a factor of *e* (2.78...) would take nearly a billion years. 1,800 degrees is higher than the actual observed exosphere temperature; at the actual average exosphere temperature, depletion of O+ ions would not occur even over a trillion years. Furthermore, most oxygen on Earth is bound as O_2, which is too massive to escape Earth by Jeans Escape.

Earth's magnetic field protects it from solar winds and prevents escape of ions, except along open field lines at the magnetic poles. The gravitational attraction of Earth's mass prevents other non-thermal loss processes from appreciably depleting the atmosphere. Yet Earth's atmosphere is two orders of magnitude less dense than that of Venus at the surface. Because of the temperature regime of Earth, CO_2 and H_2O are sequestered in the hydrosphere and lithosphere. H_2O vapour is sequestered as liquid H_2O in oceans, greatly decreasing the atmospheric density. With liquid water running over the surface of Earth, CO_2 can be drawn down from the atmosphere and sequestered in sedimentary rocks. Some estimates indicate that nearly all carbon on Earth is contained in sedimentary rocks, with the atmospheric portion being approximately 1/250,000 of Earth's CO_2 reservoir. If both of the reservoirs were released to the atmosphere, Earth's atmosphere would be denser than even Venus's atmosphere. Therefore, the dominant "loss" mechanism of Earth's atmosphere is not escape to space, but sequestration.

Water Resources

Water resources are sources of water that are useful or potentially useful. Uses of water include agricultural, industrial, household, recreational and environmental activities. Virtually all of these human uses require fresh water.

97 per cent of the water on the Earth is salt water, and only 3 per cent is fresh water of which slightly over two thirds is frozen in glaciers and polar ice caps. The remaining unfrozen freshwater is mainly found as groundwater, with only a small fraction present above ground or in the air.

Fresh water is a renewable resource, yet the world's supply of clean, fresh water is steadily decreasing. Water demand already exceeds supply in many parts of the world and as the world population continues to rise, so too does the water demand. Awareness of the global importance of preserving water for ecosystem services has only recently emerged as, during the 20th century, more than half the world's wetlands have been lost along with their valuable environmental services. Biodiversity-rich freshwater ecosystems are currently declining faster than marine or land ecosystems. The framework for allocating water resources to water users (where such a framework exists) is known as water rights.

Surface Water

Surface water is water collecting on the ground or in a stream, river, lake, wetland, or ocean; it is related to water collecting as groundwater or atmospheric water.

Surface water is naturally replenished by precipitation and naturally lost through discharge to evaporation and sub-surface seepage into the ground. Although there are other sources of groundwater, such as connate water and magmatic water, precipitation is the major one and groundwater originated in this way is called meteoric water.

Classification of Surface Water Quality

The field of hydrometry is used to characterise surface water quality:

- *Class 1* is extra clean fresh surface water resource used for conservation, not necessarily required to pass through water treatment process, and requiring only an ordinary process for pathogenic destruction and ecosystem conservation where basic organisms can breed naturally.
- *Class 2* is very clean fresh surface water resource used for consumption, which requires ordinary water treatment process before use, for aquatic organism of conservation, fisheries, and recreation.
- *Class 3* is medium clean fresh surface water resource used for consumption, but requires passing through an ordinary treatment process before use, for agriculture.
- *Class 4* is fairly clean fresh surface water resource used for consumption, but requires a special water treatment process before use, for industry.
- *Class 5* is the source which is not classified in class 1-4 and used only for navigation.

Definition: Surface water is taken from the lakes, rivers, waterfalls and sea.

Conjunctive Use of Ground and Surface Water

Surface and ground water are two separate entities, so they must be regarded as such. However, there is an ever-increasing need for management of the two as they are part of an interrelated system that is paramount when the demand for water exceeds the available supply (Fetter 464). Depletion of surface and ground water sources for public consumption (including industrial, commercial, and residential) is caused by over-pumping. Aquifers near river systems that are over-pumped have been known to deplete surface water sources as well. Research supporting this has been found in numerous water budgets for a multitude of cities. Response times for an aquifer are long (Young & Bredehoeft 1972). However, a total ban on ground water usage during water recessions would allow surface water to better retain levels required for sustainable aquatic life. By reducing ground water pumping, the surface water supplies will be able to maintain their levels, as they recharge from direct precipitation, runoff, etc.

Under River Flow

Throughout the course of a river, the total volume of water transported downstream will often be a combination of the visible free water flow together with a substantial contribution flowing through sub-surface rocks and gravels that underlie the river and its floodplain called the hyporheic zone. For many rivers in large valleys, this unseen component of flow may greatly exceed the visible flow. The hyporheic zone often forms a dynamic interface between surface water and true groundwater receiving water from the ground water when aquifers are fully charged and contributing water to groundwater when ground waters are depleted. This is especially significant in karst areas where pot-holes and underground rivers are common.

Groundwater

Groundwater is water located beneath the ground surface in soil pore spaces and in the fractures of rock formations. A unit of rock or an unconsolidated deposit is called an aquifer when it can yield a usable quantity of water. The depth at which soil pore spaces or fractures and voids in rock become completely saturated with water is called the water table. Groundwater is recharged from, and eventually flows to, the surface naturally; natural discharge often occurs at springs and seeps, and can form oases or wetlands. Groundwater is also often withdrawn for agricultural, municipal and industrial use by constructing and operating extraction wells. The study of the distribution and movement of groundwater is hydrogeology, also called groundwater hydrology.

Typically, groundwater is thought of as liquid water flowing through shallow aquifers, but technically it can also include soil moisture, permafrost (frozen soil), immobile water in very low permeability bedrock, and deep geothermal or oil formation water. Groundwater is hypothesised to provide lubrication that can possibly influence the movement of faults. It is likely that much of the Earth's subsurface contains some water, which may be mixed with other fluids in some instances. Groundwater may not be confined only to the Earth. The formation of some of the landforms observed on Mars may have been influenced by groundwater. There is also evidence that liquid water may also exist in the subsurface of Jupiter's moon Europa.

Aquifer

An aquifer is a wet underground layer of water-bearing permeable rock or unconsolidated materials (gravel, sand, or silt) from which groundwater can be usefully extracted using a water well. The study of water flow in aquifers and the characterisation of aquifers is called hydrogeology. Related terms include aquitard, which is a bed of low permeability along an aquifer, and aquiclude (or *aquifuge*), which is a solid, impermeable area underlying or overlying an aquifer. If the impermeable area overlies the aquifer pressure could cause it to become a confined aquifer.

Aquifer Depth

Aquifers may occur at various depths. Those closer to the surface are not only more likely to be used for water supply and irrigation, but are also more likely to be topped up by the local rainfall. Many desert areas have limestone hills or mountains within them or close to them that can be exploited as groundwater resources.

Parts of the Atlas Mountains in North Africa, the Lebanon and Anti-Lebanon ranges of Syria, Israel and Lebanon, the Jebel Akhdar (Oman) in Oman, parts of the Sierra Nevada and neighbouring ranges in the United States' Southwest, have shallow aquifers that are exploited for their water. Over-exploitation can lead to the exceeding of the practical sustained yield; i.e., more water is taken out than can be replenished.

Along the coastlines of certain countries, such as Libya and Israel, population growth has led to over-population, which has caused the lowering of water table and the subsequent contamination of the groundwater with saltwater from the sea (saline intrusions).

The beach provides a model to help visualise an aquifer. If a hole is dug into the sand, very wet or saturated sand will be located at a shallow depth. This hole is a crude well, the wet sand represents an aquifer, and the level to which the water rises in this hole represents the water table.

Classification

This diagram indicates typical flow directions in a cross-sectional view of a simple confined/unconfined aquifer system. The system shows two aquifers with one aquitard (a confining or impermeable layer), between them, surrounded by the bedrock *aquiclude*, which is in contact with a gaining stream (typical in humid regions). The water table and unsaturated zone are also illustrated. An *aquitard* is a zone within the earth that restricts the flow of groundwater from one aquifer to another. An aquitard can sometimes, if completely impermeable, be called an *aquiclude* or *aquifuge*. Aquitards are composed of layers of either clay or non-porous rock with low hydraulic conductivity.

Saturated versus Unsaturated

Groundwater can be found at nearly every point in the Earth's shallow subsurface, to some degree; although aquifers do not necessarily contain fresh water. The Earth's crust can be divided into two regions: the *saturated zone* or *phreatic zone* (e.g., aquifers, aquitards, etc.), where all available spaces are filled with water, and the *unsaturated zone* (also called the vadose zone), where there are still pockets of air with some water, but can be filled with more water.

Saturated means the pressure head of the water is greater than atmospheric pressure (it has a gauge pressure > 0). The definition of the water table is surface where the pressure head is equal to atmospheric pressure (where gauge pressure = 0).

Unsaturated conditions occur above the water table where the pressure head is negative (absolute pressure can never be negative, but gauge pressure can) and the water that incompletely fills the pores of the aquifer material is under suction. The water content Unsaturated means the zone is held in place by surface adhesive forces and it rises above the water table (the zero gauge pressure isobar) by capillary action to saturate a small zone above the phreatic surface (the capillary fringe) at less than atmospheric pressure. This is termed tension saturation and is not the same as saturation on a water content basis. Water content in a capillary fringe decreases with increasing distance from the phreatic surface. The capillary head depends on soil pore size. In sandy soils with larger pores, the head will be less than in clay soils with very small pores. The normal capillary rise in a clayey soil is less than 1.80 m (six feet) but can range between 0.3 and 10 m (1 and 30 ft).

The capillary rise of water in a small diameter tube is this same physical process. The water table is the level to which water will rise in a large-diameter pipe (e.g., a well) that goes down into the aquifer and is open to the atmosphere.

Aquifers versus Aquitards

Aquifers are typically saturated regions of the subsurface that produce an economically feasible quantity of water to a well or spring (e.g., sand and gravel or fractured bedrock often make good aquifer materials).

An aquitard is a zone within the earth that restricts the flow of groundwater from one aquifer to another. An aquitard can sometimes, if completely impermeable, be called an aquiclude or aquifuge. Aquitards comprise layers of either clay or non-porous rock with low hydraulic conductivity.

In mountainous areas (or near rivers in mountainous areas), the main aquifers are typically unconsolidated alluvium, composed of mostly horizontal layers of materials deposited by water processes (rivers and streams), which in cross-section (looking at a two-dimensional slice of the aquifer) appear to be layers of alternating coarse and fine materials. Coarse materials, because of the high energy needed to move them, tend to be found nearer the source (mountain fronts or rivers), whereas the fine-grained material will make it farther from the source (to the flatter parts of the basin or overbank areas - sometimes called the pressure area). Since there are less fine-grained deposits near the source, this is a place where aquifers are often unconfined (sometimes called the forebay area), or in hydraulic communication with the land surface.

Confined versus Unconfined

There are two end members in the spectrum of types of aquifers; *confined* and *unconfined* (with semi-confined being in between). Unconfined aquifers are sometimes also called *water table* or *phreatic* aquifers, because their upper boundary is the water table or phreatic surface. Typically (but not always) the shallowest aquifer at a given location is unconfined, meaning it does not have a confining layer (an aquitard or aquiclude) between it and the surface. The term "perched" refers to ground water accumulating above a low-permeability unit or strata, such as a clay layer. This term is generally used to refer to a small local area of ground water that occurs at an elevation higher than a regionally-extensive aquifer. The difference between perched and unconfined aquifers is their size (perched is smaller).

If the distinction between confined and unconfined is not clear geologically (i.e., if it is not known if a clear confining layer exists, or if the geology is more complex, e.g., a fractured bedrock aquifer), the value of storativity returned from an aquifer test can be used to determine it (although aquifer tests in unconfined aquifers should be interpreted differently than confined ones). Confined aquifers have very low storativity values (much less than 0.01, and as little as 10^{-5}), which means that the aquifer is storing water using the mechanisms of aquifer matrix expansion and the compressibility of water, which typically are both quite small quantities. Unconfined aquifers have storativities (typically then called specific yield) greater than 0.01 (1% of bulk volume); they release water from storage by the mechanism of actually draining the pores of the aquifer, releasing relatively large amounts of water (up to the drainable porosity of the aquifer material, or the minimum volumetric water content).

Isotropic versus Anisotropic

In isotropic aquifers or aquifer layers the hydraulic conductivity (K) is equal for flow in all directions, while in anisotropic conditions it differs, notably in horizontal (Kh) and vertical (Kv) sense.

Semi-confined aquifers with one or more aquitards work as an anisotropic system, even when the separate layers are isotropic, because the compound Kh and Kv values are different.

When calculating flow to drains or flow to wells in an aquifer, the anisotropy is to be taken into account lest the resulting design of the drainage system may be faulty.

Groundwater in Rock Formations

Groundwater may exist in *underground rivers* (e.g., caves where water flows freely underground). This may occur in eroded limestone areas known as karst topography, which make up only a small percentage of Earth's area. More usual is that the pore spaces of rocks in the subsurface are simply saturated with water — like a kitchen sponge — which can be pumped out for agricultural, industrial, or municipal uses.

If a rock unit of low porosity is highly fractured, it can also make a good aquifer (via fissure flow), provided the rock has an appreciable hydraulic conductivity to facilitate movement of water. Porosity is important, but, *alone*, it does not determine a rock's ability of being an aquifer. Areas of the Deccan Traps (a basaltic lava) in west central India are good examples of rock formations with high porosity but low permeability, which makes them poor aquifers. Similarly, the micro-porous (Upper Cretaceous) Chalk of south east England, although having a reasonably high porosity, has a low grain-to-grain permeability, with much of its good water-yielding characteristics being due to micro-fracturing and fissuring.

Human Dependence on Groundwater

Most land areas on Earth have some form of aquifer underlying them, sometimes at significant depths.

Fresh-water aquifers, especially those with limited recharge by meteoric water, can be over-exploited and, depending on the local hydrogeology, may draw in non-potable water or saltwater (saltwater intrusion) from hydraulically connected aquifers or surface water bodies. This can be a serious problem, especially in coastal areas and other areas where aquifer pumping is excessive.

Aquifers are critically important in human habitation and agriculture. Deep aquifers in arid areas have long been water sources for irrigation. Many villages and even large cities draw their water supply from wells in aquifers.

Municipal, irrigation, and industrial water supplies are provided through large wells. Multiple wells for one water supply source are termed "wellfields", which may withdraw water from confined or unconfined aquifers. Using ground water from deep, confined aquifers provides more protection from surface water contamination. Some wells, termed "collector wells," are specifically designed to induce infiltration of surface (usually river) water.

Aquifers that provide sustainable fresh groundwater to urban areas and for agricultural irrigation are typically close to the ground surface (within a couple of hundred metres) and have some recharge by fresh water. This recharge is typically from rivers or meteoric water (precipitation) that percolates into the aquifer through overlying unsaturated materials. Aquifer depletion has been cited as one of the causes of the Great Food Crisis of 2011.

Subsidence

In unconsolidated aquifers, groundwater is produced from pore spaces between particles of gravel, sand, and silt. If the aquifer is confined by low-permeability layers, the reduced water pressure in the sand and gravel causes slow drainage of water from the adjoining confining layers. If these confining layers are composed of compressible silt or clay, the loss of water to the aquifer reduces the water pressure in the confining layer, causing it to compress from the weight of overlying geologic materials. In severe cases, this compression can be observed on the ground surface as subsidence. Unfortunately, much of the subsidence from groundwater extraction is permanent (elastic rebound is small). Thus, the subsidence is not only permanent, but the compressed aquifer has a permanently-reduced capacity to hold water.

Saltwater Intrusion

Aquifers near the coast have a lens of freshwater near the surface and denser seawater under freshwater. Seawater penetrates the aquifer diffusing in from the ocean and is denser than freshwater. For porous (i.e., sandy) aquifers near the coast, the thickness of freshwater atop saltwater is about 40 feet (12 m) for every 1 ft (0.30 m) of freshwater head above sea level. This relationship is called the Ghyben-Herzberg equation. If too much ground water is pumped near the coast, salt-water may intrude into freshwater aquifers causing contamination of potable freshwater supplies. Many coastal aquifers, such as the Biscayne Aquifer near Miami and the New Jersey Coastal Plain aquifer, have problems with saltwater intrusion as a result of overpumping.

Salination

Aquifers in surface irrigated areas in semi-arid zones with reuse of the unavoidable irrigation water losses percolating down into the underground by supplemental irrigation from wells run the risk of salination.

Surface irrigation water normally contains salts in the order of 0.5 g/l or more and the annual irrigation requirement is in the order of 10000 m/ha or more so the annual import of salt is in the order of 5000 kg/ha or more.

Under the influence of continuous evaporation, the salt concentration of the aquifer water may increase continually and eventually cause an environmental problem.

For salinity control in such a case, annually an amount of drainage water is to be discharged from the aquifer by means of a subsurface drainage system and disposed of through a safe outlet. The drainage system may be *horizontal* (i.e. using pipes, tile drains or ditches) or *vertical* (drainage by wells). To estimate the drainage requirement, the use of a groundwater model with an agro-hydro-salinity component may be instrumental, e.g. SahysMod.

Examples of Aquifers

The Great Artesian Basin situated in Australia is arguably the largest groundwater aquifer in the world (over 1.7 million km^2). It plays a large part in water supplies for Queensland and remote parts of South Australia.

The Guarani Aquifer with an area of 1.2 million km^2 is shared by Brazil, Argentina, Paraguay and Uruguay. Aquifer depletion is a problem in some areas, and is especially critical in northern Africa; see the Great Manmade River project of Libya for an example. However, new methods of groundwater management such as artificial recharge and injection of surface waters during seasonal wet periods has extended the life of many freshwater aquifers, especially in the United States. The Ogallala Aquifer of the central United States is one of the world's great aquifers, but in places it is being rapidly depleted by growing municipal use, and continuing agricultural use. This huge aquifer, which underlies portions of eight states, contains primarily fossil water from the time of the last glaciation. Annual recharge, in the more arid parts of the aquifer, is estimated to total only about 10 per cent of annual withdrawals. An example of a significant and sustainable carbonate aquifer is the Edwards Aquifer in central Texas. This carbonate aquifer has historically been providing high quality water for nearly 2 million people, and even today, is completely full because of tremendous recharge from a number of area streams, rivers and lakes. The primary risk to this resource is human development over the recharge areas.

Climatic Effects of Aquifer Depletion

Aquifer drawdown or overdrafting and the pumping of fossil water increases the total amount of water in the hydrosphere that is subject to transpiration and evaporation thereby causing accretion in water vapour and cloud cover which are the primary absorbers of infrared radiation in the earth's atmosphere. Adding water to the system has a forcing effect on the whole earth system, an accurate estimate of which hydrogeological fact is yet to be quantified.

Issues

Overview

Certain problems have beset the use of groundwater around the world. Just as river waters have been over-used and polluted in many parts of the world, so too have aquifers. The big difference is that aquifers are out of sight. The other major problem is that water management agencies, when calculating the 'sustainable yield' of aquifer and river water, have often counted the same water twice, once in the aquifer, and once in its connected river. This problem, although understood for centuries, has persisted, partly through inertia within government agencies. In Australia, for example, prior to the statutory reforms initiated by the Council of Australian Governments water reform framework in the 1990s, many Australian States managed groundwater and surface water through separate government agencies, an approach beset by rivalry and poor communication. The time lags inherent in the dynamic response of groundwater to development have generally been ignored by water management agencies, decades after scientific understanding of the issue was consolidated. In brief, the effects of groundwater overdraft (although undeniably real) may take decades or centuries to manifest themselves. In a classic study in 1982, Bredehoeft and colleagues modelled a situation where groundwater extraction in an intermontane basin withdrew the entire annual recharge, leaving 'nothing' for the natural groundwater-dependent vegetation community. Even when the borefield was situated close to the vegetation, 30 per cent of the original vegetation demand could still be met by the lag inherent in the system after 100 years. By year 500 this had reduced to 0 per cent, signalling complete death of the groundwater-dependent vegetation. The science has been available to make these calculations for decades; however water management agencies have generally ignored effects which will appear outside the rough timeframe of political elections (3 to 5 years). Marios Sophocleous argued strongly that management agencies must define and use appropriate timeframes in groundwater planning. This will mean calculating groundwater withdrawal permits based on predicted effects decades, sometimes centuries in the future.

As water moves through the landscape it collects soluble salts, mainly sodium chloride. Where such water enters the atmosphere through evapotranspiration, these salts are left behind. In irrigation districts, poor drainage of soils and surface aquifers can result in water tables coming to the surface in low-lying areas. Major land degradation problems of soil salinity and waterlogging result, combined with increasing levels of salt in surface waters. As a consequence, major damage has occurred to local economies and environments. Four important effects are worthy of brief mention. First, flood mitigation schemes, intended to protect infrastructure built on floodplains, have had the unintended consequence of reducing aquifer recharge associated with natural flooding. Second, prolonged depletion of groundwater in extensive aquifers can result in land subsidence, with associated infrastructure damage – as well as (thirdly) saline intrusion. Fourth, draining acid sulphate soils, often found in low-lying coastal plains, can result in acidification and pollution of formerly freshwater and estuarine streams. Another cause for concern is that groundwater drawdown from over-allocated aquifers has the potential to cause severe damage to both terrestrial and aquatic ecosystems – in some cases very conspicuously but in others quite imperceptibly because of the extended period over which the damage occurs.

Overdrafting

Overdrafting is the process of extracting groundwater beyond the *safe yield* or equilibrium yield of the aquifer.

Since every groundwater basin recharges at a different rate depending upon precipitation, vegetative cover and soil conservation practises, the quantity of groundwater that can be safely pumped varies greatly among regions of the world and even within provinces. Some aquifers require a very long time to recharge and thus the process of overdrafting can have consequences of effectively drying up certain sub-surface water supplies.

Effects on Climate

Aquifer drawdown or overdrafting and the pumping of fossil water increases the total amount of water within the hydrosphere subject to transpiration and evaporation processes, thereby causing accretion in water vapour and cloud cover, the primary absorbers of infrared radiation in the earth's atmosphere. Adding water to the system has a forcing effect on the whole earth system, an accurate estimate of which hydrogeological fact is yet to be quantified.

Socio-economic Effects

Scores of countries are overpumping aquifers as they struggle to satisfy their growing water needs, including each of the big three grain producers—China, India, and the United States. These three, along with a number of other countries where water tables are falling, are home to more than half the world's people. Water is an intrinsic to growth . Overdraft limits the availability of supply.

10

Groundwater-related Subsidence

Groundwater-related subsidence is the subsidence (or the sinking) of land resulting from groundwater extraction, and a major problem in the developing world as major metropolises swell without adequate regulation and enforcement, as well as a being a common problem in the developed world. One estimate has 80 per cent of serious land subsidence problems associated with the excessive extraction of groundwater, making it a growing problem throughout the world.

Groundwater is considered to be one of the last 'free' resources. Anybody who can afford to drill, can draw up merely according to the their ability to pump. However, the act of pumping draws down the free surface of the groundwater table, and can affect a large region. Thus, the extraction of groundwater becomes a Tragedy of the commons, with high economic externalities. The desert areas of the world are requiring more and more water for growing populations, and agriculture. In the San Joaquin Valley of the United States, groundwater pumping for crops has gone on for generations. This has resulted in the entire valley sinking an extraordinary amount, as shown in the figure. Fortunately, there is little consequence in a wide, flat, agricultural basin, since the settlement is uniform. Total settlement can only be determined by surveys and GPS measurements.

Unfortunately, not all groundwater-related subsidence is benign. In Mexico City the buildings interact with the settlement, and cause cracking, tilting, and other major damage. In many places, large sinkholes open up, as well as surface cavities. Damage from Hurricane Katrina was exacerbated due to coastal sinking, associated with groundwater withdrawal.

The cause of the long-term surface changes associated with this phenomenon, are fairly well known.

The only known method to prevent this condition is by pumping less groundwater, which is extremely difficult to enforce when many people own water wells. Attempts are being made to directly recharge aquifers but this still a preliminary effort.

Seawater Intrusion

Generally, in very humid or undeveloped regions, the shape of the water table mimics the slope of the surface. The recharge zone of an aquifer near the seacoast is likely to be inland, often at considerable distance. In these coastal areas, a lowered water table may induce sea water to reverse the flow towards the land. Sea water moving inland is called a saltwater intrusion. Alternatively, salt from mineral beds may leach into the groundwater of its own accord.

Mining

Sometimes the water movement from the recharge zone to the place where it is withdrawn may take centuries. When the usage of water is greater than the recharge, it is referred to as *mining* water (the water is often called fossil water because of its geologic age). Under those circumstances it is not a renewable resource.

Ground Water Rule

In November 2006, the Environmental Protection Agency published the Ground Water Rule in the United States Federal Register. The EPA was worried that the ground water system would be vulnerable to contamination from fecal matter. The point of the rule was to keep microbial pathogens out of public water sources The 2006 Ground Water Rule was an amendment of the 1996 Safe Drinking Water Act.

Desalination

Desalination, desalinisation, or desalinisation refers to any of several processes that remove some amount of salt and other minerals from water. More generally, desalination may also refer to the removal of salts and minerals, as in soil desalination.

Water is desalinated in order to convert salt water to fresh water so it is suitable for human consumption or irrigation. Sometimes the process produces table salt as a by-product. Desalination is used on many seagoing ships and submarines. Most of the modern interest in desalination is focused on developing cost-effective ways of providing fresh water for human use in regions where the availability of fresh water is, or is becoming, limited.

Large-scale desalination typically uses extremely large amounts of energy as well as specialised, expensive infrastructure, making it very costly compared to the use of fresh water from rivers or groundwater. However, along with recycled water this is one of the few non-rainfall dependent water sources particularly relevant to countries like Australia which traditionally have relied on rainfall in dams to provide their drinking water supplies.

The world's largest desalination plant is the Jebel Ali Desalination Plant (Phase 2) in the United Arab Emirates. It is a dual-purpose facility that uses multi-stage flash distillation and is capable of producing 300 million cubic metres of water per year. By comparison the largest desalination plant in the United States is located in Tampa Bay, Florida, and operated by Tampa Bay Water, which began desalinating 34.7 million cubic meters of water per year in December 2007.

The Tampa Bay plant runs at around 12 per cent the output of the Jebel Ali Desalination Plants. The largest desalination plant in South Asia is the Minjur Desalination Plant near Chennai in India which produces 100,000 cubic meters of water per day, or 36.5 million cubic meters of water per year. According to International Desalination Association 2009, there are 14,451 desalination plants in operation worldwide, producing 59.9 million cubic meters per day (15.8 billion gallons a day), a year on year increase of 12.3 per cent.

Methods

The traditional process used in these operations is vacuum distillation—essentially the boiling of water at less than atmospheric pressure and thus a much lower temperature than normal. This is because the boiling of a liquid occurs when the vapour pressure equals the ambient pressure and vapour pressure increases with temperature.

Thus, because of the reduced temperature, energy is saved. A leading distillation method is multi-stage flash distillation accounting for 85 per cent of production worldwide in 2004.

The principal competing processes use membranes to desalinate, principally applying reverse osmosis technology. Membrane processes use semi-permeable membranes and pressure to separate salts from water. Reverse osmosis plant membrane systems typically use less energy than thermal distillation, which has led to a reduction in overall desalination costs over the past decade. Desalination remains energy intensive, however, and future costs will continue to depend on the price of both energy and desalination technology.

Considerations and Criticism

Cogeneration

Cogeneration is the process of using excess heat from power production to accomplish another task. For desalination, cogeneration is the production of potable water from seawater or brackish groundwater in an integrated, or "dual-purpose", facility in which a power plant is used as the source of energy for the desalination process. The facility's energy production may be dedicated entirely to the production of potable water (a stand-alone facility), or excess energy may be produced and incorporated into the energy grid (a true cogeneration facility).

There are various forms of cogeneration, and theoretically any form of energy production could be used. However, the majority of current and planned cogeneration desalination plants use either fossil fuels or nuclear power as their source of energy. Most plants are located in the Middle East or North Africa, due to their petroleum resources and subsidies. The advantage of dual-purpose facilities is that they can be more efficient in energy consumption, thus making desalination a more viable option for drinking water in areas of scarce water resources.

In a December 26, 2007, opinion column in the *The Atlanta Journal-Constitution*, Nolan Hertel, a professor of nuclear and radiological engineering at Georgia Tech, wrote, "... nuclear reactors can be used ... to produce large amounts of potable water. The process is already in use in a number of places around the world, from India to Japan and Russia.

Eight nuclear reactors coupled to desalination plants are operating in Japan alone ... nuclear desalination plants could be a source of large amounts of potable water transported by pipelines hundreds of miles inland..." Additionally, the current trend in dual-purpose facilities is hybrid configurations, in which the permeate from an RO desalination component is mixed with distillate from thermal desalination. Basically, two or more desalination processes are combined along with power production. Such facilities have already been implemented in Saudi Arabia at Jeddah and Yanbu. A typical aircraft carrier in the U.S. military uses nuclear power to desalinate 400,000 US gallons (1,500,000 l; 330,000 imp gal) of water per day.

Economics

A number of factors determine the capital and operating costs for desalination: capacity and type of facility, location, feed water, labor, energy, financing, and concentrate disposal. Desalination stills now control pressure, temperature and brine concentrations to optimise the water extraction efficiency. Nuclear-powered desalination might be economical on a large scale.

While noting that costs are falling, and generally positive about the technology for affluent areas that are proximate to oceans, one study argues that "Desalinated water may be a solution for some water-stress regions, but not for places that are poor, deep in the interior of a continent, or at high elevation. Unfortunately, that includes some of the places with biggest water problems." and "Indeed, one needs to lift the water by 2,000 metres (6,600 ft), or transport it over more than 1,600 kilometres (990 mi) to get transport costs equal to the desalination costs.

Thus, it may be more economical to transport fresh water from somewhere else than to desalinate it. In places far from the sea, like New Delhi, or in high places, like Mexico City, high transport costs would add to the high desalination costs. Desalinated water is also expensive in places that are both somewhat far from the sea and somewhat high, such as Riyadh and Harare.

In many places, the dominant cost is desalination, not transport; the process would therefore be relatively less expensive in places like Beijing, Bangkok, Zaragoza, Phoenix, and, of course, coastal cities like Tripoli." After being desalinated at Jubail, Saudi Arabia, water is pumped 200 miles (320 km) inland through a pipeline to the capital city of Riyadh. For cities on the coast, desalination is being increasingly viewed as an untapped and unlimited water source.

Desalination makes sense only after less expensive options are exhausted, including recycling water and fixing broken infrastructure. Water is reused in Las Vegas NV, Fountain Valley CA, Fairfax VA, El Paso TX and Scottsdale AZ. Compared to desalinated sea water, recycling requires 50 per cent less energy due to the significantly lower salt content and produces new water at 30 per cent less cost to the consumer, without the damage to marine life and ecosystems common to desalination plants.

Israel is now desalinating water at a cost of US$0.53 per cubic meter. Singapore is desalinating water for US$0.49 per cubic meter. Many large coastal cities in developed countries are considering the feasibility of seawater desalination, due to its cost effectiveness compared with other water supply options, which can include mandatory installation of rainwater tanks or stormwater harvesting infrastructure.

Studies have shown that the desalination option is more cost-effective than large-scale recycled water for drinking, and more cost-effective in Sydney than the vastly expensive option of mandatory installation of rainwater tanks or stormwater harvesting infrastructure. The city of Perth has been successfully operating a reverse osmosis seawater desalination plant since 2006, and the Western Australian government have announced that a second plant will be built to serve the city's needs. A desalination plant is now operating in Australia's largest city of Sydney, and the Wonthaggi desalination plant under construction in Wonthaggi, Victoria.

The Perth desalination plant is powered partially by renewable energy from the Emu Downs Wind Farm. A wind farm at Bungendore in NSW has been purpose-built to generate enough renewable energy to offset the energy use of the Sydney plant, mitigating concerns about harmful greenhouse gas emissions, a common argument used against seawater desalination due to the energy requirements of the technology.

The purchase or production of renewable energy to power desalination plants naturally adds to the capital and/or operating costs of desalination. However, recent experience in Perth and Sydney indicates that the additional cost is acceptable to communities, as a city may then augment its water supply without doing environmental harm to the atmosphere.

The Queensland state government also purchased renewable energy certificates on behalf of its Gold Coast plant which will see the plant offset its carbon emissions for the initial 18 to 20 months of operations, bringing its environmental footprint down, in line with the other major plants that will be operating around the same time, in Perth and Sydney.

In December 2007, the South Australian government announced that it would build a seawater desalination plant for the city of Adelaide, Australia, located at Port Stanvac. The desalination plant is to be funded by raising water rates to achieve full cost recovery.

An online, unscientific poll showed that nearly 60 per cent of votes cast were in favour of raising water rates to pay for desalination. A January 17, 2008, article in the *Wall Street Journal* states, "In November, Connecticut-based Poseidon Resources Corp. won a key regulatory approval to build the US$300 million water-desalination plant in Carlsbad, north of San Diego.

The facility would produce 50,000,000 US gallons (190,000,000 l; 42,000,000 imp gal) of drinking water per day, enough to supply about 100,000 homes ... Improved technology has cut the cost of desalination in half in the past decade, making it more competitive ... Poseidon plans to sell the water for about US $950 per acre-foot [1,200 cubic metres (42,000 cu ft)]. That compares with an average US$700 an acre-foot [1200 m^3] that local agencies now pay for water." $1,000 per acre-foot works out to $3.06 for 1,000 gallons, or $.81 for 1 cubic meter, which is the unit of water measurement that residential water users are accustomed to being billed in.

While this regulatory hurdle was met, Poseidon Resources is not able to break ground until the final approval of a mitigation project for the damage done to marine life through the intake pipe, as is required by California law.

Poseidon Resources has made progress in Carlsbad, CA, despite its unsuccessful attempt to complete construction of Tampa Bay Desal, a desalination plant in Tampa Bay, FL, in 2001. The Board of Directors of Tampa Bay Water were forced to buy Tampa Bay Desal from Poseidon Resources in 2001 to prevent a third failure of the project. Tampa Bay Water faced five years of engineering problems and operation at 20 per cent capacity due to marine life and growth captured and stuck to reverse osmosis filters prior to fully utilising this facility in 2007.

According to a May 9, 2008, article in *Forbes*, a San Leandro, California, company called Energy Recovery Inc. has been desalinating water for US $0.46 per cubic meter. According to a June 5, 2008, article in the *Globe and Mail*, a Jordanian-born chemical engineering doctoral student at the University of Ottawa, named Mohammed Rasool Qtaisha, has invented a new desalination technology that is alleged to be between 600 per cent and 700 per cent more water output per square meter of membrane than current technology.

According to the article, General Electric is looking into similar technology, and the U.S. National Science Foundation announced a grant to the University of Michigan to study it as well. Because the patents were still being worked out, the article was very vague about the details of this alleged technology. While desalinating 1,000 US gallons (3,800 l; 830 imp gal) of water can cost as much as $3, the same amount of bottled water costs $7,945.

Environmental

Intake: One of the main environmental considerations of ocean water desalination plants is the impact of the open ocean water intakes, especially when co-located with power plants. Many proposed ocean desalination plants' initial plans relied on these intakes despite perpetuating ongoing impacts on marine life. In the United States, due to a recent court ruling under the Clean Water Act, these intakes are no longer viable without reducing mortality, by 90 per cent, of the life in the ocean; the plankton, fish eggs and fish larvae. There are alternatives, including beach wells that eliminate this concern, but require more energy and higher costs while limiting output. Other environmental concerns include air pollution and greenhouse gas emissions from the power plants.

Outflow: To limit the environmental impact of returning the brine to the ocean, it can be diluted with another stream of water entering the ocean, such as the outfall of a waste water treatment plant or power plant. While seawater power plant cooling water outfalls are not freshwater like waste water treatment plant outfalls, the salinity of the brine will still be reduced. If the power plant is medium- to large-sized and the desalination plant is not enormous, the flow of the power plant's cooling water is likely to be at least several times larger than that of the desalination plant. Another method to reduce the increase in salinity is to spread the brine via a diffuser to mix in a mixing zone so that there is only a slight increase in salinity. For example, once the pipeline containing the brine reaches the sea floor, it can split off into many branches, each one releasing the brine gradually along its length through small holes. This method can be used in combination with the joining of the brine with power plant or waste water plant outfalls.

There are methods of desalination, particularly in combination with open pond evaporation (solar desalination), that do not discharge brine back into the ocean at all. The concentrated seawater has the potential to harm ecosystems, especially marine environments in regions with low turbidity and high evaporation that already have elevated salinity. Examples of such locations are the Persian Gulf, the Red Sea and, in particular, coral lagoons of atolls and other tropical islands around the world.

The UAE, Qatar, Bahrain, Saudi Arabia, Kuwait and Iran have 120 desalination plants between them. These plants flush nearly 24 tons of chlorine, 65 tons of algae-harming antiscalants used to descale pipes, and around 300 kg of copper into the Persian Gulf every day. Because the brine is denser than the surrounding sea water due to the higher solute concentration, discharge into water bodies means that the ecosystems on the bed of the water body are most at risk because the brine sinks and remains there long enough to damage the ecosystems.

Careful re-introduction can minimise this problem. For example, for the desalination plant and ocean outlet structures to be built in Sydney from late 2007, the water authority states that the ocean outlets will be placed in locations at the seabed that will maximise the dispersal of the concentrated seawater, such that it will be indistinguishable from normal seawater between 50 and 75 metres (160 and 246 ft) from the outlet points. Sydney is fortunate to have typical oceanographic conditions off the coast that allow for such rapid dilution of the concentrated by-product, thereby minimising harm to the environment.

In Perth, Australia, in 2007, the Kwinana Desalination Plant was opened. The water is sucked in from the ocean at only 0.1 metres per second (0.33 ft/s), which is slow enough to let fish escape. The plant provides nearly 140,000 cubic metres (4,900,000 cu ft) of clean water per day. This is the same at Queensland's Gold Coast Desalination Plant and Sydney's Desalination Plant.

Desalination Compared to other Water Supply Options

Increased water conservation and water use efficiency remain the most cost-effective priorities in areas of the world where there is a large potential to improve the efficiency of water use practices. While comparing ocean water desalination to waste water reclamation for drinking water shows desalination as the first option, using reclamation for irrigation and industrial use provides multiple benefits. Urban runoff and storm water capture also provide benefits in treating, restoring and recharging groundwater. A proposed alternative to desalinisation in the state of California and other areas in the American Southwest is the commercial importation of bulk water either by very large crude carriers converted to water carriers, or via pipelines.

Experimental Techniques and other Developments

In the past, many novel desalination techniques have been researched with varying degrees of success. One such process which has recently been commercialised by Modern Water plc is a forward osmosis based process for desalinated water, with a number of plants reported in operation. Other techniques have also attracted research funding. For example, to offset the energy requirements of desalination, the U.S. government is working to develop practical solar desalination. As an example of newer theoretical approaches for desalination, focusing specifically on maximising energy efficiency and cost effectiveness, the Passarell Process may be considered.

Other approaches involve the use of geothermal energy. From an environmental and economic point of view, in most locations geothermal desalination can be preferable to using fossil groundwater or surface water for human needs, as in many regions the available surface and groundwater resources already have long been under severe stress. Recent research in the U.S. indicates that nanotube membranes may prove to be extremely effective for water filtration and may produce a viable water desalination process that would require substantially less energy than reverse osmosis.

Another method being looked into for water desalination is the use of biomimetic membranes. On June 23, 2008, it was reported that Siemens Water Technologies had developed a new technology, based on applying electric field on seawater, that desalinates one cubic meter of water while using only 1.5 kWh of energy, which, according to the report, is one half the energy that other processes use.

Fresh water can also be produced by freezing seawater, as happens naturally in the polar regions, and is known as freeze-thaw desalination. According to MSNBC, a report by Lux Research estimated that the worldwide desalinated water supply will triple between 2008 and 2020.

Desalination through Evaporation and Condensation for Crops Growth

The Seawater Greenhouse uses evaporation and condensation natural processes inside a greenhouse powered by solar energy to enable the growth of crops in coastal arid land.

Low-temperature Thermal Desalination

Low-temperature thermal desalination (LTTD) takes advantage of the fact that water boils at low pressures, even as low as ambient temperature. The system uses vacuum pumps to create a low pressure, low-temperature environment in which water boils at a temperature gradient of 8 to 10°C between two volumes of water. Cooling water is supplied from sea depths of as much as 600 metres (2,000 ft).

This cold water is pumped through coils to condense the evaporated water vapour. The resulting condensate is purified water. The LTTD process may also take advantage of the temperature gradient available at power plants, where large quantities of warm waste water are discharged from the plant, reducing the energy input needed to create a temperature gradient.

The principle of LTTD is known for a long time, originally stemming from ocean thermal energy conversion research. Some experiments were conducted in U.S. and Japan to test the low-temperature driven desalination technology. In Japan, a spray flash evaporation system was tested by Saga University. In US, at Hawaii Islands, the National Energy Laboratory tested an open-cycle OTEC plant with fresh water and power production using a temperature of 20°C between surface water and water at a depth of around 500 m.

LTTD was studied by India's National Institute of Ocean Technology (NIOT) from 2004. Their first LTTD plant was opened in 2005 at Kavaratti in the Lakshadweep islands. The plant's capacity is 100,000 litres (22,000 imp gal; 26,000 US gal)/day, at a capital cost of INR 50 million (€922,000). The plant uses deep water at a temperature of 7 to 15°C (45 to 59 °F). In 2007, NIOT opened an experimental floating LTTD plant off the coast of Chennai with a capacity of 1,000,000 litres (220,000 imp gal; 260,000 US gal)/day. A smaller plant was established in 2009 at the North Chennai Thermal Power Station to prove the LTTD application where power plant cooling water is available.

Thermo-ionic Process

In October 2009, Saltworks Technologies, a Canadian firm, announced a process that uses solar or other thermal heat to drive an ionic current that empties all the sodium and chlorine ions from the water.

Existing Facilities and Facilities under Construction

Aruba

The island of Aruba has a large (world's largest at the time of its inauguration) desalination plant with the total installed capacity of 42,000 metric tons (11.1 million gallons or 42×10^3 m^3) per day.

Australia

A combination of increased water usage and lower rainfall/drought in Australia has caused State governments to build a number of desalination plants, including the recently commissioned Kurnell Desalination Plant serving the Sydney area. While desalination has been adopted by state governments to secure water supply, it is highly energy intensive (~$140 energy demand/ ML) and has a high carbon footprint due to continued reliance on Australia's coal-based energy generation.

Bahrain

- The Al Hidd Desalination Plant on Muharraq island treats seawater through a multistage flash process, and produces 30 million US gallons (110,000 m^3) per day. This project was completed in 2000. The Al Hidd distillate forwarding station, comprises a 410 million litres distillate water storage in 45 million litres steel tanks. A 135 million litres/day forwarding pumping station sends flows to the Hidd blending station, Muharraq blending station, Hoora blending station, Sanabis blending station and Seef blending station and which has an option for gravity supply for low flows to blending pumps and pumps which forward to Janusan, Budiya and Saar.
- When completed in three phases, the Durrat Al Bahrain sea water reverse osmosis (SWRO) desalination plant will have a capacity of 36,000 cubic meters of potable water per day which will serve the irrigation needs of the entire Durrat Al Bahrain development. The Bahrain-based utility company, Energy Central Co (ECC) will provide the plant a 25-year design, build and operate contract.

China

China operates the Beijiang Desalination Plant in Tianjin, a combination desalination and coal-fired power plant designed to alleviate Tianjin's critical water shortage. Though the facility has the capacity to produce 200,000 cubic meters of potable water per day, it has never operated at more than one quarter capacity due to difficulties with local utility companies and an inadequate local infrastructure.

Cyprus

There are also desalination plants in Cyprus, like the one near the town of Larnaca. This is called the Dhekelia Desalination Plant, which utilises the reverse osmosis system.

Gibraltar

The fresh water supply in Gibraltar is supplied by a number of reverse osmosis and multi-stage flash desalination plants. There is also a demonstration forward osmosis desalination plant operational.

Israel

The Hadera seawater reverse osmosis (SWRO) desalination plant in Israel is the largest of its kind in the world. The project was developed as a Build-Operate-Transfer (BOT) by a consortium of three international companies: Veolia water, IDE Technologies and Elran.

***Table 1:** Existing Israeli Water Desalination Facilities*

Location	*Opened*	*Capacity (mln m^3/year)*	*Cost of water (per m^3)*
Ashkelon	August 2005	120 (as of 2010)	NIS 2.60
Palmachim	May 2007	45	NIS 2.90
Hadera	December 2009	127	NIS 2.60

***Table 2:** Israeli water desalination facilities under construction*

Location	*Opening*	*Capacity (mln m^3/year)*	*Cost of water (per m^3)*
Ashdod	2012	100 (expansion up to 150 possible)	NIS 2.40
Soreq	2013	150 (expansion up to 300 approved)	NIS 2.01 – 2.19

Maldives

Maldives is a small island nation and most of the islands depend on desalination as a source of water.

Oman

A pilot Seawater Greenhouse was built in 2004 near Muscat, Oman. In collaboration with Sultan Qaboos University, providing an opportunity to develop a sustainable horticultural sector on the Batinah coast .

Saudi Arabia

The Saline Water Conversion Corporation of Saudi Arabia provides 50 per cent of the municipal water in the Kingdom, operates a number of desalination plants, and has contracted $1892 million to a Japanese-South Korean consortium to build one capable of producing a billion litres a day, opening at the end of 2013. They currently operate approximately 14 plants in the Kingdom; one example at Shoaiba cost $1060 million and produces 450 million litres a day.

Spain

Lanzarote, Canary Islands: Lanzarote is the easternmost of the autonomous Canary Islands. It is of volcanic origin and has only very limited natural water supplies. A commercial desalination plant was installed in 1964 as a private initiative. This served the whole island and started off the tourism industry. In 1974 the venture was injected with investments from local and municipal governments and a larger infrastructure was put in place. In 1989 INALSA was formed as the Lanzarote Island Waters Consortium. The plant has been inspected by foreign visitors from places with interest in desalination.

Tenerife, Canary Islands

A prototype Seawater Greenhouse was constructed in Tenerife in 1992 .

El Prat del Llobregat, Catalonia

El Prat, near Barcelona, has a desalination plant completed in 2009 and meant to provide water to the Barcelona metropolitan area, specially during the periodic severe droughs that put the drinking water provision under serious stress.

United Arab Emirates (Abu Dhabi)

- Taweelah A1 Power and Desalination Plant has an output 385,000,000 litres (85,000,000 imp gal; 102,000,000 US gal) per day of clean water.
- Umm Al Nar Desalination Plant has an output of 394,000,000 litres (87,000,000 imp gal; 104,000,000 US gal) per day of clean water.
- Fujairah F2 is to be completed by July 2010 will have a water production capacity of 492,000,000 litres (108,000,000 imp gal; 130,000,000 US gal) per day.
- A Seawater Greenhouse was constructed on Al-Aryam Island, Abu Dhabi, United Arab Emirates in 2000.

United Kingdom

Beckton Desalination Plant: The first large scale water desalination plant in the United Kingdom, the Thames Water Desalination Plant, has been built in Beckton, east London for Thames Water by Acciona Agua

United States

El Paso (Texas) Desalination Plant: Brackish groundwater has been treated at the El Paso plant since around 2004. Producing 27,500,000 US gallons (104,000,000 l; 22,900,000 imp gal) of fresh water daily (about 25% of total freshwater deliveries) by reverse osmosis, it is a crucial contribution to water supplies in this water-stressed city.

Tampa Bay Water Desalination Project

The Tampa Bay Water Desalination project was originally a private venture led by Poseidon Resources. This project was delayed by the bankruptcy of Poseidon Resources' successive partners in the venture, Stone & Webster, then Covanta (formerly Ogden) and its principal subcontractor Hydranautics.

Poseidon's relationship with Stone & Webster through S & W Water LLC ended in June 2000 when Stone & Webster declared bankruptcy and Poseidon Resources purchased Stone & Webster's stake in S & W Water LLC. Poseidon Resources partnered with Covanta and Hydranautics in 2001, changing the consortium name to Tampa Bay Desal.

Through the inability of Covanta to complete construction bonding of the project, the Tampa Bay Water agency was forced to purchase the project from Poseidon on May 15, 2002, and underwrite the project financing under its own credit rating. Tampa Bay Water then contracted with Covanta Tampa Construction, which produced a project that did not meet required performance tests. Covanta Tampa Construction's parent company filed bankruptcy in October 2003 to prevent losing the contract with Tampa Bay Water.

Then, Covanta Tampa Construction filed bankruptcy prior to performing renovations that would have satisfied contractual agreements. This resulted in nearly six months of litigation between Covanta Tampa Construction and Tampa Bay Water. In 2004, Tampa Bay Water hired a renovation team, American Water/Acciona Aqua, to bring the plant to its original, anticipated design.

The plant was deemed fully operational in 2007 and is designed to run at a maximum capacity of 25 million US gallons (95,000 m^3) per day. Nevertheless, the plant continues to be set with problems limiting it to producing only about half that amount (14 million US gallons (53,000 m^3) per day in 2009.

Yuma Desalting Plant (Arizona)

The Yuma Desalting Plant was constructed under authority of the Colorado River Basin Salinity Control Act of 1974 to treat saline agricultural return flows from the Wellton-Mohawk Irrigation and Drainage District. The treated water is intended for inclusion in water deliveries to Mexico thereby preserving the like amount of water in Lake Mead.

Construction of the plant was completed in 1992 and it has operated on two occasions since then. The plant has been maintained, but largely not operated due to surplus and then normal water supply conditions on the Colorado River. An agreement was reached in April 2010 between the Southern Nevada Water Authority, the Metropolitan Water District of Southern California, the Central Arizona Project and the U.S. Bureau of Reclamation to underwrite the cost of running the plant in a year long pilot project.

Trinidad and Tobago

The Republic of Trinidad and Tobago is using desalination to free up more of the island's water supply for drinking purposes. The desalination facility, opened in March 2003, is considered to be the first of its kind. It is the largest desalination facility in the Americas and will process 28,800,000 US gallons (109,000,000 l; 24,000,000 imp gal) of water a day and sell water at the price of $2.67 per 1,000 US gallons (3,800 l; 830 imp gal). This facility will be located at Trinidad's Point Lisas Industrial Estate, a park of more than 12 companies in various manufacturing and processing functions and will allow for easy access to water for both factories and residents in the country.

In Nature

Seabirds distill sea water, by using Countercurrent exchange in a gland with a Rete mirabile. The gland secretes highly concentrated brine stored near the nostrils above the beak. The bird then "sneezes" the brine out to the sea. As freshwater is not available in their environment, seabirds like pelicans, petrels, albatrosses, gulls and terns possess this gland which allows the birds to drink the salty water from their environment while they are hundreds miles away from land.

Mangroves are trees which grow in sea water. Mangroves are able to secrete the salt by trapping it into parts of the root, which are then eaten by animals (usually crabs). Additional salt removal is done by storing it in leaves which then fall off. Some types of mangrove have glands on the leaf, which work in a similar way to the seabird desalination gland.

Salt is extracted to the leaf exterior as small crystals, which then fall off the leaf. Willow trees and Reeds are known to uptake salt and other contaminants, effectively desalinating the water. This is used in artificial constructed wetlands, for treating sewage.

Rain and Dew are the outcome of large scale desalination through evaporation, in what is known as the water cycle. Fresh water floats above sea water, so during storms or in lakes where there is undersea volcanic activity causing the water to boil and then condensate, large bodies of freshwater can be found at the sea or lake surface, covering the more salty seawater below. Freshwater on the surface of seawater is common at estuaries.

Frozen Water

Several schemes have been proposed to make use of icebergs as a water source, however to date this has only been done for novelty purposes. Glacier runoff is considered to be surface water.

The Himalayas, which are often called "The Roof of the World", contain some of the most extensive and rough high altitude areas on Earth as well as the greatest area of glaciers and permafrost outside of the poles. Ten of Asia's largest rivers flow from there, and more than a billion people's livelihoods depend on them. To complicate matters, temperatures are rising more rapidly here than the global average. In Nepal the temperature has risen with 0.6 degree over the last decade, whereas the global warming has been around 0.7 over the last hundred years.

Uses of Fresh Water

Uses of fresh water can be categorised as consumptive and non-consumptive (sometimes called "renewable"). A use of water is consumptive if that water is not immediately available for another use. Losses to sub-surface seepage and evaporation are considered consumptive, as is water incorporated into a product (such as farm produce).

Water that can be treated and returned as surface water, such as sewage, is generally considered non-consumptive if that water can be put to additional use. Water use in power generation and industry is generally described using an alternate terminology, focusing on separate measurements of withdrawal and consumption. Withdrawal describes the removal of water from the environment, while consumption describes the conversion of fresh water into some other form, such as atmospheric water vapour or contaminated waste water

Agricultural

It is estimated that 69 per cent of worldwide water use is for irrigation, with 15-35 per cent of irrigation withdrawals being unsustainable. It takes around 3,000 litres of water, converted from liquid to vapour, to produce enough food to satisfy one person's daily dietary need. This is a considerable amount, when compared to that required for drinking, which is between two and five litres. To produce food for the 6.5 billion or so people who inhabit the planet today requires the water that would fill a canal ten metres deep, 100 metres wide and 7.1 million kilometres long – that's enough to circle the globe 180 times.

Increasing Water Scarcity

Fifty years ago, the common perception was that water was an infinite resource. At this time, there were fewer than half the current number of people on the planet. People were not as wealthy as today, consumed fewer calories and ate less meat, so less water was needed to produce their food. They required a third of the volume of water we presently take from rivers.

Today, the competition for water resources is much more intense. This is because there are now nearly seven billion people on the planet, their consumption of water-thirsty meat and vegetables is rising, and there is increasing competition for water from industry, urbanisation and biofuel crops. In future, even more water will be needed to produce food because the Earth's population is forecast to rise to 9 billion by 2050. An additional 2.5 or 3 billion people, choosing to eat fewer cereals and more meat and vegetables could add an additional five million kilometres to the virtual canal mentioned above.

An assessment of water management in agriculture was conducted in 2007 by the International Water Management Institute in Sri Lanka to see if the world had sufficient water to provide food for its growing population. It assessed the current availability of water for agriculture on a global scale and mapped out locations suffering from water scarcity.

It found that a fifth of the world's people, more than 1.2 billion, live in areas of physical water scarcity, where there is not enough water to meet all demands. A further 1.6 billion people live in areas experiencing economic water scarcity, where the lack of investment in water or insufficient human capacity make it impossible for authorities to satisfy the demand for water. The report found that it would be possible to produce the food required in future, but that continuation of today's food production and environmental trends would lead to crises in many parts of the world.

To avoid a global water crisis, farmers will have to strive to increase productivity to meet growing demands for food, while industry and cities find ways to use water more efficiently. In some areas of the world irrigation is necessary to grow any crop at all, in other areas it permits more profitable crops to be grown or enhances crop yield. Various irrigation methods involve different trade-offs between crop yield, water consumption and capital cost of equipment and structures.

Irrigation methods such as furrow and overhead sprinkler irrigation are usually less expensive but are also typically less efficient, because much of the water evaporates, runs off or drains below the root zone. Other irrigation methods considered to be more efficient include drip or trickle irrigation, surge irrigation, and some types of sprinkler systems where the sprinklers are operated near ground level.

These types of systems, while more expensive, usually offer greater potential to minimise runoff, drainage and evaporation. Any system that is improperly managed can be wasteful, all methods have the potential for high efficiencies under suitable conditions, appropriate irrigation timing and management. Some issues that are often insufficiently considered are salinisation of sub-surface water and contaminant accumulation leading to water quality declines.

As global populations grow, and as demand for food increases in a world with a fixed water supply, there are efforts under way to learn how to produce more food with less water, through improvements in irrigation methods and technologies, agricultural water management, crop types, and water monitoring.

Aquaculture is a small but growing agricultural use of water. Freshwater commercial fisheries may also be considered as agricultural uses of water, but have generally been assigned a lower priority than irrigation.

Industrial

It is estimated that 22 per cent of worldwide water use is industrial. Major industrial users include hydroelectric dams, thermoelectric power plants, which use water for cooling, ore and oil refineries, which use water in chemical processes, and manufacturing plants, which use water as a solvent.

Water withdrawal can be very high for certain industries, but consumption is generally much lower than that of agriculture. Water is used in renewable power generation. Hydroelectric power derives energy from the force of water flowing downhill, driving a turbine connected to a generator. This hydroelectricity is a low-cost, non-polluting, renewable energy source.

Significantly, hydroelectric power can also be used for load following unlike most renewable energy sources which are intermittent. Ultimately, the energy in a hydroelectric powerplant is supplied by the sun. Heat from the sun evaporates water, which condenses as rain in higher altitudes and flows downhill. Pumped-storage hydroelectric plants also exist, which use grid electricity to pump water uphill when demand is low, and use to stored water to produce electricity when demand is high.

Hydroelectric power plants generally require the creation of a large artificial lake. Evaporation from this lake is higher than evaporation from a river due to the larger surface area exposed to the elements, resulting in much higher water consumption. The process of driving water through the turbine and tunnels or pipes also briefly removes this water from the natural environment, creating water withdrawal.

The impact of this withdrawal on wildlife varies greatly depending on the design of the powerplant. Pressurised water is used in water blasting and water jet cutters. Also, very high pressure water guns are used for precise cutting. It works very well, is relatively safe, and is not harmful to the environment. It is also used in the cooling of machinery to prevent over-heating, or prevent saw blades from over-heating. This is generally a very small source of water consumption relative to other uses.

Water is also used in many large scale industrial processes, such as thermoelectric power production, oil refining, fertilizer production and other chemical plant use, and natural gas extraction from shale rock. Discharge of untreated water from industrial uses is pollution. Pollution includes discharged solutes (chemical pollution) and increased water temperature (thermal pollution). Industry requires pure water for many applications and utilises a variety of purification techniques both in water supply and discharge. Most of this pure water is generated on site, either from natural freshwater or from municipal grey water.

Industrial consumption of water is generally much lower than withdrawal, due to laws requiring industrial grey water to be treated and returned to the environment. Thermoelectric powerplants using cooling towers have high consumption, nearly equal to their withdrawal, as most of the withdrawn water is evaporated as part of the cooling process. The withdrawal, however, is lower than in once-through cooling systems.

Household

It is estimated that 8 per cent of worldwide water use is for household purposes. These include drinking water, bathing, cooking, sanitation, and gardening. Basic household water requirements have been estimated by Peter Gleick at around 50 litres per person per day, excluding water for gardens. Drinking water is water that is of sufficiently high quality so that it can be consumed or used without risk of immediate or long term harm.

Such water is commonly called potable water. In most developed countries, the water supplied to households, commerce and industry is all of drinking water standard even though only a very small proportion is actually consumed or used in food preparation.

Recreation

Recreational water use is usually a very small but growing percentage of total water use. Recreational water use is mostly tied to reservoirs. If a reservoir is kept fuller than it would otherwise be for recreation, then the water retained could be categorised as recreational usage. Release of water from a few reservoirs is also timed to enhance whitewater boating, which also could be considered a recreational usage.

Other examples are anglers, water skiers, nature enthusiasts and swimmers. Recreational usage is usually non-consumptive. Golf courses are often targeted as using excessive amounts of water, especially in drier regions. It is, however, unclear whether recreational irrigation (which would include private gardens) has a noticeable effect on water resources. This is largely due to the unavailability of reliable data.

Additionally, many golf courses utilise either primarily or exclusively treated effluent water, which has little impact on potable water availability. Some governments, including the Californian Government, have labelled golf course usage as agricultural in order to deflect environmentalists' charges of wasting water. However, the actual statistical effect of this reassignment is close to zero. In Arizona, an organised lobby has been established in the form of the Golf Industry Association, a group focused on educating the public on how golf impacts the environment. Recreational usage may reduce the availability of water for other users at specific times and places. For example, water retained in a reservoir to allow boating in the late summer is not available to farmers during the spring planting season. Water released for whitewater rafting may not be available for hydroelectric generation during the time of peak electrical demand.

Environmental

Explicit environmental water use is also a very small but growing percentage of total water use. Environmental water may include water stored in impoundments and released for environmental purposes (held environmental water), but more often is water retained in waterways through regulatory limits of abstraction.

Environmental water usage includes watering of natural or artificial wetlands, artificial lakes intended to create wildlife habitat, fish ladders, and water releases from reservoirs timed to help fish spawn, or to restore more natural flow regimes. Like recreational usage, environmental usage is non-consumptive but may reduce the availability of water for other users at specific times and places.

For example, water release from a reservoir to help fish spawn may not be available to farms upstream, and water retained in a river to maintain waterway health would not be available to water abstractors downstream.

Water Stress

Researchers define water stress and water scarcity in different ways. For example, some have presented maps showing the physical existence of water in nature to show nations with lower or higher volumes of water available for use.

Others have related water availability to population. A popular approach has been to rank countries according to the amount of annual water resources available per person.

For example, according to the Falkenmark Water Stress Indicator, a country or region is said to experience "water stress" when annual water supplies drop below 1,700 cubic metres per person per year. At levels between 1,700 and 1,000 cubic metres per person per year, periodic or limited water shortages can be expected.

When water supplies drop below 1,000 cubic metres per person per year, the country faces "water scarcity". The Unitied Nations' FAO states that by 2025, 1.8 billion people will be living in countries or regions with absolute water scarcity, and two-thirds of the world population could be under stress conditions. Water stress is considered "set to become a permanent feature of

British life". Another measurement, calculated as part of a wider assessment of water management in 2007, aimed to relate water availability to how the resource was actually used.

It therefore divided water scarcity into 'physical' and 'economic'. Physical water scarcity is where there is not enough water to meet all demands, including that needed for ecosystems to function effectively. Arid regions frequently suffer from physical water scarcity.

It also occurs where water seems abundant but where resources are over-committed, such as when there is overdevelopment of hydraulic infrastructure for irrigation. Symptoms of physical water scarcity include environmental degradation and declining groundwater.

Economic water scarcity, meanwhile, is caused by a lack of investment in water or insufficient human capacity to satisfy the demand for water. Symptoms of economic water scarcity include a lack of infrastructure, with people often having to fetch water from rivers for domestic and agricultural uses.

Large parts of Africa suffer from economic water scarcity; developing water infrastructure there could therefore help to reduce poverty. Critical conditions often arise for economically poor and politically weak communities living in already dry environments. Some 2.8 billion people currently live in water-scarce areas, as defined by this method.

Causes of Water Stress

The concepts of water stress and water scarcity are relatively new. Fifty years ago, when there were fewer than half the current number of people on the planet, the common perception was that water was an infinite resource. People were not as wealthy then as they are today, consumed fewer calories and ate less meat, so less water was needed to produce their food.

They required a third of the volume of water we presently take from rivers. Today, the competition for water resources is much more intense. This is because there are now nearly seven billion people on the planet, their consumption of water-thirsty meat and vegetables is rising, and there is increasing competition for water from industry, urbanisation and biofuel crops.

The total amount of available freshwater supply is also decreasing because of climate change, which has caused receding glaciers, reduced stream and river flow, and shrinking lakes. Many aquifers have been over-pumped and are not recharging quickly.

Although the total fresh water supply is not used up, much has become polluted, salted, unsuitable or otherwise unavailable for drinking, industry and agriculture. To avoid a global water crisis, farmers will have to strive to increase productivity to meet growing demands for food, while industry and cities find ways to use water more efficiently.

Water Right

Water right in water law refers to the right of a user to use water from a water source, e.g., a river, stream, pond or source of groundwater. In areas with plentiful water and few users, such systems are generally not complicated or contentious. In other areas, especially arid areas where irrigation is practiced, such systems are often the source of conflict, both legal and physical. Some systems treat surface water and ground water in the same manner, while others use different principles for each.

Types of Water Rights

There are fundamental differences between the nature and source of water rights in different countries, including that they can be land-based rights, use-based rights, or rights based on ownership of water bodies. In addition, some countries apply the principle of water solidarity, which has become increasingly important in the EU. This principle attempts to reconcile water rights through compromise solutions based on the public interest, and it has been written into legislation in France, for example.

Land-based or Riparian Rights

Riparian rights are based on land ownership, and are protected by property law. Riparian rights state that only the owner of the banks of the water source have a right to the 'undiminished, unaltered flow' of the water. Riparian rights are only transferable when the riparian land ownership title is transferred to a new owner.

Use-based Rights

Use-based water rights are protected by the law of torts. Use-based rights state that land ownership is not essential, as long as water users have legal access to the water source. There is a hierarchy of use, where the first user has the strongest rights (first in, first served). Rights users can only enforce rights against users with lower ranks (those who came later). Use-based rights are usufructuary, fully transferable to anyone.

Water Rights Based on Ownership of Water Bodies

In Finland, for example, waterbodies are generally privately owned, which is not the case in most EU countries, but Finland also applies the Roman law principle of *aqua profluens* (flowing water), according to which the freely flowing water in waterbodies cannot be owned or possessed. This means that the owners of waterbodies cannot prohibit diversion of water for agricultural, industrial, municipal, or domestic use according to the provisions of the Finnish Water Law and cannot prohibit use of the waterbodies for recreational purposes.

History of Water Rights

In Roman times, the law was that people could obtain temporary usufructuary rights for running water. These rights were independent of land ownership, and lasted as long as use continued. Under Roman law, no land was "owned" by citizens, it was all owned by the "republic" and controlled by politicians.

In Medieval times, the common law of the day treated an all freshwater streams as static, meaning landowners owned parts of rivers, with full accompanying rights. Landowners could also seek damages for loss of water diverted upstream. Non-landowners did not have use rights, except by obtaining a prescription. Over time, rights evolved from being land based to use based, allowing non-landowners to hold enforceable rights. A reasonable use theory evolved in some countries.

Water Rights in the United States

In the United States, there are two divergent systems for determining water rights. Riparian water rights (derived from English common law) are common in the east and prior appropriation water rights (developed in Colorado and California) are common in the west.

Each state has its own variations on these basic principles, as informed by custom, culture, geography, legislation and case law. California law, for example, includes elements of both systems. In general, a water right is established by obtaining an authorisation from the state in the form of a water right permit. A legal right is formally consummated, or perfected, by exercising the water right permit and using the water for a beneficial purpose.

Under the prior appropriation doctrine, water rights are "first in time, first in right." That is, the older, or senior, water right may operate to the exclusion of junior water rights. The concept of "priority date" is significant. The priority date is generally associated with the date that water was first put to beneficial use, or the date that a successful application for a water right was submitted, and indicates the relative status of seniority among competing users.

Older rights are senior. More recent rights are junior. Water rights are generally established pursuant to State law, but there are exceptions, most notably, the concept of federal reserved water rights. Reserved water rights are rights that are established when the federal government reserves land for a specific federal purpose. Courts have held that there is an implied water right to satisfy the primary purposes of the reservation. Examples of reservations include Indian reservations, national wildlife refuges, federal forests and military bases.

Proceedings to determine the relative priority of claims to water rights are known as adjudications. Through Congress's passage of the McCarren amendment, the federal government has consented to having its claims adjudicated in state courts.

All states offer mechanisms for changing how a water right is exercised, e.g., amending the point of diversion or withdrawal, the place of use and the purpose of use. In reviewing such requests, the state must guard against the impairment of other water rights, the enlargement of the water right and injury to the public interest.

Resolution of Interstate Water Conflicts

Because water bodies may cross political and jurisdictional boundaries, conflicts may arise. In the United States, three basic approaches are used to settle such conflicts: 1) Litigation before the Supreme Court of the United States; 2) Legislative resolution by the Congress of the United States; and 3) Negotiation and ratification of interstate compacts between states.

In the western United States, for example, the 1922 Colorado River Compact divides the Colorado River basin into two areas, the Upper Division (comprising Colorado, New Mexico, Utah and Wyoming) and the Lower Division (Nevada, Arizona and California). A comprehensive review of existing interstate water compacts has been published by the Model Interstate Water Compact Project at the University of New Mexico School of Law's Utton Transboundary Resource Centre.

Limitations on Water Rights

In California, courts have held that appropriation water rights may be limited under the public trust doctrine, a common law principle recognised by some courts, which holds that the public has access rights upon navigable waters and that navigable waters are held in trust for the use of the people.

The public trust doctrine was invoked by the California Supreme Court in a case restricting the amount of water Los Angeles could divert from tributaries of Mono Lake. The case was filed by the Audubon Society and the Mono Lake Committee.

In the United States, Navigable waters are subject to the commerce clause of the U.S. Constitution. The commerce clause provides the federal government the ability to restrict state issued water rights via, for example, the enforcement of water quality standards via the Federal Water Pollution Control Act (Clean Water Act), the Federal Power Act and the protection of endangered species via the Endangered Species Act.

Fresh Water

Fresh water is naturally occurring water on the Earth's surface in ice sheets, ice caps, glaciers, bogs, ponds, lakes, rivers and streams, and underground as groundwater in aquifers and underground streams. Fresh water is generally characterised by having low concentrations of dissolved salts and other total dissolved solids. The term specifically excludes seawater and brackish water although it does include mineral rich waters such as chalybeate springs. The term "sweet water" has been used to describe fresh water in contrast to salt water.

Systems

Scientifically, freshwater habitats are divided into lentic systems, which are the stillwaters including ponds, lakes, swamps and mires; lotic systems, which are running water; and groundwater which flows in rocks and aquifers. There is, in addition, a zone which bridges between groundwater and lotic systems, which is the hyporheic zone, which underlies many larger rivers and can contain substantially more water than is seen in the open channel. It may also be in direct contact with the underlying underground water.

Source

The source of almost all fresh water is precipitation from the atmosphere, in the form of mist, rain and snow. Fresh water falling as mist, rain or snow contains materials dissolved from the atmosphere and material from the sea and land over which the rain bearing clouds have traveled. In industrialised areas rain is typically acidic because of dissolved oxides of sulphur and nitrogen formed from burning of fossil fuels in cars, factories, trains and aircraft and from the atmospheric emissions of industry.

In extreme cases this acid rain results in pollution of lakes and rivers in parts of Scandinavia, Scotland, Wales and the United States. In coastal areas fresh water may contain significant concentrations of salts derived from the sea if windy conditions have lifted drops of seawater into the rain-bearing clouds.

This can give rise to elevated concentrations of sodium, chloride, magnesium and sulphate as well as many other compounds in smaller concentrations. In desert areas, or areas with impoverished or dusty soils, rain-bearing winds can pick up sand and dust and this can be deposited elsewhere in precipitation and causing the freshwater flow to be measurably contaminated both by insoluble solids but also by the soluble components of those soils. Significant quantities of iron may be transported in this way including the well-documented transfer of iron-rich rainfall falling in Brazil derived from sand-storms in the Sahara in north Africa.

Numerical Definition

Fresh water can be defined as water with less than 500 parts per million (ppm) of dissolved salts.

Table 2: Water salinity based on dissolved salts in parts per thousand (‰)

Fresh water	*Brackish water*	*Saline water*	*Brine*
< 0.5	0.5 – 30	30 – 50	> 50

Other sources give higher upper salinity limits for fresh water, e.g. 1000 ppm or 3000 ppm.

Water Distribution

Water is a critical issue for the survival of all living organisms. Some can use salt water but many organisms including the great majority of higher plants and most mammals must have access to fresh water to live. Some terrestrial mammals, especially desert rodents appear to survive without drinking but they do generate water through the metabolism of cereal seeds and they also have mechanisms to conserve water to the maximum degree.

Out of all the water on Earth, only 2.75 per cent is fresh water, including 2.05 per cent frozen in glaciers, 0.68 per cent as groundwater and 0.011 per cent of it as surface water in lakes and rivers. Freshwater lakes, most notably Lake Baikal in Russia and the Great Lakes in North America, contain seven-eighths of this fresh surface water.

Swamps have most of the balance with only a small amount in rivers, most notably the Amazon River. The atmosphere contains 0.04 per cent water. In areas with no fresh water on the ground surface, fresh water derived from precipitation may, because of its lower density, overlie saline ground water in lenses or layers. Most of the world's fresh water is frozen in ice sheets. Many areas suffer from lack of distribution of fresh water, such as deserts and uncommonly known; Florida, US.

Aquatic Organisms

Fresh water creates a hypotonic environment for aquatic organisms. This is problematic for some organisms with pervious skins or with gill membranes, whose cell membranes may burst if excess water is not excreted. Some protists accomplish this using contractile vacuoles, while freshwater fish excrete excess water via the kidney. Although most aquatic organisms have a limited ability to regulate their osmotic balance and therefore can only live within a narrow range of salinity, diadromous fish have the ability to migrate between fresh water and saline water bodies.

During these migrations they undergo changes to adapt to the surroundings of the changed salinities; these processes are hormonally controlled. The eel (*Anguilla anguilla*) uses the hormone prolactin, while in salmon (*Salmo salar*) the hormone cortisol plays a key role during this process. Many sea birds have special glands at the base of the bill through which excess salt is excreted. Similarly the Marine Iguanas on the Galapagos Islands excrete excess salt through a nasal gland and they sneeze out a very salty excretion.

Fresh Water as a Resource

An important concern for hydrological ecosystems is securing minimum streamflow, especially preserving and restoring instream water allocations. Fresh water is an important natural resource necessary for the survival of all ecosystems.

The use of water by humans for activities such as irrigation and industrial applications can have adverse impacts on down-stream ecosystems. Chemical contamination of fresh water can also seriously damage eco-systems. Pollution from human activity, including oil spills, also presents a problem for freshwater resources. The largest oil spill that has ever occurred in fresh water was caused by a Shell tank ship in Magdalena, Argentina, on January 15, 1999, polluting the environment, drinkable water, plants and animals. Fresh and unpolluted water accounts for 0.003 per cent of total water available globally.

Agriculture

Changing landscape for the use of agriculture has a great effect on the flow of fresh water. Changes in landscape by the removal of trees and soils changes the flow of fresh water in the local environment and also affects the cycle of fresh water.

As a result more fresh water is stored in the soil which benefits agriculture. However, since agriculture is the human activity that consumes the most fresh water, this can put a severe strain on local freshwater resources resulting in the destruction of local ecosystems. In Australia, over-abstraction of fresh water for intensive irrigation activities has caused 33 per cent of the land area to be at risk of salination.

Limiting Resource

Fresh water is a renewable and changeable, but limited natural resource. Fresh water can only be renewed through the process of the water cycle, where water from seas, lakes, rivers, and dams evaporates, forms clouds, and returns to water sources as precipitation. However, if more fresh water is consumed through human activities than is restored by nature, the result is that the quantity of fresh water available in lakes, rivers, dams and underground waters is reduced which can cause serious damage to the surrounding environment.

Fresh Water Withdrawal

Fresh water withdrawal is the quantity of water removed from available sources for use in any purpose. Water drawn-off is not necessarily entirely consumed and some portion may be returned for further use downstream.

Causes of Limited Fresh Water

There are many causes of the apparent decrease in our fresh water supply. Principal amongst these is the increase in population through increasing life expectancy, the increase in per capita water use and the desire of many people to live in warm climates that have naturally low levels of fresh water resources. Climate change is also likely to change the availability and distribution of fresh water across the planet:

"If global warming continues to melt glaciers in the polar regions, as expected, the supply of fresh water may actually decrease. First, fresh water from the melting glaciers will mingle with salt water in the oceans and become too salty to drink. Second, the increased ocean volume will cause sea levels to rise, contaminating freshwater sources along coastal regions with seawater". Water pollution and subsequent eutrophication also reduces the availability of fresh water.

Seawater

Seawater is water from a sea or ocean. On average, seawater in the world's oceans has a salinity of about 3.5 per cent (35 g/L, or 599 mM). This means that every kilogram (roughly one litre by volume) of seawater has approximately 35 grams (1.2 oz) of dissolved salts (predominantly sodium (Na^+) and chloride (Cl^-) ions).

The average density of seawater at the ocean surface is 1.025 g/ml. Seawater is denser than both fresh water and pure water (density 1.0 g/ml @ 4°C (39 °F)) because the dissolved salts add mass without contributing significantly to the volume. The freezing point of seawater decreases as salt concentration increases. At a typical salinity it freezes at about "2°C (28 °F). The coldest seawater ever recorded (in a liquid state) was in 2010, in a stream under an Antarctic glacier, and measured "2.6°C (27.3 °F).

Salinity

Salinity is the saltiness or dissolved salt content of a body of water. It is a general term used to describe the levels of different salts such as sodium chloride, magnesium and calcium sulphates, and bicarbonates. *Salinity* in Australian English and North American English may also refer to the salt content of soil.

Definitions

Table 3: *Water salinity*

Fresh water	*Brackish water*	*Saline water*	*Brine*
< 0.05 %	0.05 – 3 %	3 – 5 %	> 5 %
< 0.5 ‰	0.5 – 30 ‰	30 – 50 ‰	> 50 ‰

Salinity in the ocean refers to the water's "saltiness". In oceanography, it has been traditional to express salinity not as per cent, but as parts per thousand (‰), which is approximately grams of salt per kilogram of solution. Other disciplines use chemical analyses of solutions, and thus salinity is frequently reported in mg/L or ppm (parts per million).

Prior to 1978, salinity or halinity was expressed as ‰ usually based on the electrical conductivity ratio of the sample to "Copenhagen water", an artificial sea water manufactured to serve as a world standard. In 1978, oceanographers redefined salinity in the Practical Salinity Scale (PSS) as the conductivity ratio of a sea water sample to a standard KCl solution. As PSS is a ratio, it has no units. It is not the case that a salinity of 35 exactly equals 35 grams of salt per liter of solution.

These seemingly esoteric approaches to measuring and reporting salt concentrations may appear to obscure their practical use; but it must be remembered that salinity is the sum weight of many different elements within a given volume of water.

It has always been the case that to get a precise salinity as a concentration and convert this to an amount of substance (sodium chloride, for instance) required knowing much more about the sample and the measurement than just the weight of the solids upon evaporation (one method of determining salinity). For example, volume is influenced by water temperature; and also the composition of the salts is not a constant (although generally very much the same throughout the world ocean).

Saline waters from inland seas can have a composition that differs from that of the ocean. For the latter reason, these waters are termed saline as differentiated from ocean waters, where the term haline applies (although is not universally used).

Systems of Classification of Water Bodies Based Upon Salinity

Marine waters are those of the ocean, another term for which is euhaline seas. The salinity of euhaline seas is 30 to 35. Brackish seas or waters have salinity in the range of 0.5 to 29 and metahaline seas from 36 to 40. These waters are all regarded as thalassic because their salinity is derived from the ocean and defined as homoiohaline if salinity does not vary much over time (essentially constant).

In contrast to homoiohaline environments are certain poikilohaline environments (which may also be *thallassic*) in which the salinity variation is biologically significant. *Poikilohaline* water salinities may range anywhere from 0.5 to greater than 300. The important characteristic is that these waters tend to vary in salinity over some biologically meaningful range seasonally or on some other roughly comparable time scale. Put simply, these are bodies of water with quite variable salinity.

Highly saline water, from which salts crystallise (or are about to), is referred to as brine.

Environmental Considerations

Salinity is an ecological factor of considerable importance, influencing the types of organisms that live in a body of water. As well, salinity influences the kinds of plants that will grow either in a water body, or on land fed by a water (or by a groundwater). A plant adapted to saline conditions is called a halophyte. Organisms (mostly bacteria) that can live in very salty conditions are classified as extremophiles, halophiles specifically. An organism that can withstand a wide range of salinities is euryhaline.

Salt is expensive to remove from water, and salt content is an important factor in water use (such as potability).

The degree of salinity in oceans is a driver of the world's ocean circulation, where density changes due to both salinity changes and temperature changes at the surface of the ocean produce changes in buoyancy, which cause the sinking and rising of water masses. Changes in the salinity of the oceans are thought to contribute to global changes in carbon dioxide as more saline waters are less soluble to carbon dioxide. In addition, during glacial periods, the hydrography is such that a possible cause of reduced circulation is the production of stratified oceans. Hence it is difficult in this case to subduct water through the thermohaline circulation.

Bibliography

Anderson, Mary P. and William W. Woessner: *Applied Groundwater Modeling: Simulation of Flow and Advective Transport.* Academic Press, San Diego, 1992.

Asaithambi, S. : *Economics of Ground Water Management in India*, Abhijeet, Delhi, 2008.

Bhakar, S.R. : *Ground Water Hydrology : Theory and Practice*, Agrotech, Delhi, 2009.

Bhave, P.R. and R. Gupta: *Analysis of Water Distribution Networks*, Narosa, Delhi, 2011.

Biswas Asit K. : *Integrated Water Resources Management in South and South-East Asia*, , Oxford University Press, Delhi, 2001.

Bourne, Peter G.: *Water and Sanitation: Economic and Sociological Perspectives* Academic Press. Orlando. 1984.

Brooks, D.: *Water: Local-Level Management*, Ottawa, International Development Research Centre, 2002.

Cullis, A. : *Rainwater Harvesting: The Collection of Rainfall and Runoff in Rural Areas,* London, U.K.: IT Publications, 1986.

Erik Nissen-Peterson : *Rainwater Catchment Systems*, UK: Intermediate Technology Publications, 1999.

Feder, Paul A.: *A Guidebook to Groundwater: Resources and Education Opportunities in the Great Lakes Region*, Great Lakes Commission, Ann Arbor, 1993.

Fetter, Charles Willard: *Applied Hydrogeology*, Merrill Publishing Company, Columbus, OH. 1988.

Geraghty, James J.: *Water Atlas of the United States*, Port Washington, NY, 1973.

Gischler, Christiaan E.: *Water Resources in the Arab Middle East and North Africa*, Cambridge, Eng, Middle East & North African Studies Press, 1979.

Gurjar, Ram Kumar : *Geography of Water Resources*, Rawat, Delhi, 2008.

Hemenway, Toby : *Gaia's Garden: A Guide to Home-Scale Permaculture*, Vermont: Chelsea Green Publishing Company, 2000.

Jat, M.L. and S.R. Bhakar: *Ground Water Hydrology : Theory and Practice*, Agrotech, Delhi, 2009.

Kanmony, J. Cyril : *Drinking Water Management : Problems and Prospects*, Mittal Pub, Delhi, 2010.

LeRoy, K. : *Sustaining Water Washington*, D.C.: Population Services International, New York, 1993.

McGarry, M.G.: *Matching Water Supply Technology to the Needs and Resources of Developing Countries,* United Nations. New York. 1987

Meenu Bhatnagar: *Groundwater Management : Sustainable Approaches*, Icfai Books, Delhi, 2012.

Morris, Mary E.: *Water Scarcity and Security concerns in the Middle East*, Abu Dhabi, Emirates Center for Strategic Studies and Research, 1998.

Narwani, G.S. : *Community Water Management*, Rawat, Delhi, 2005.

Nielsen, David M.: *Practical Handbook of Groundwater Monitoring*. Lewis Publishers, Chelsea, 1991.

Palanisami, K : *Groundwater Management and Policies*, Macmillan Publishers India, Delhi, 2008.

Premjit Sharma: *Agricultural Drainage and Water Quality*, Gene Tech Books, Delhi, 2007.

Singh, A.K.: *Environment and Water Resources Management*, Adhyayan, Delhi, 2006.

Thomas, Harold E.: *The Conservation of Groundwater: A Survey of the Present Groundwater Situation in the U.S.,* McGraw Hill, New York, NY. 1951.

Todd, David Keith: *Groundwater Hydrology*. John Wiley & Sons, New York, NY. 1959.

William, K. : *Irrigation Investment Technology, and Management Strategies for Development Studies in Water Policy Management 9*, Boulder, Colo Westview Press, 1986.

Index

❑❑❑